Learn, Practice, Succeed

Eureka Math®
Grade 6
Module 4

Published by Great Minds®.

Copyright © 2019 Great Minds®.

Printed in the U.S.A.

This book may be purchased from the publisher at eureka-math.org.

10 9 8 7 6 5 4 3 2

ISBN 978-1-64054-967-8

G6-M4-LPS-05.2019

Students, families, and educators:

Thank you for being part of the *Eureka Math®* community, where we celebrate the joy, wonder, and thrill of mathematics.

In *Eureka Math* classrooms, learning is activated through rich experiences and dialogue. That new knowledge is best retained when it is reinforced with intentional practice. The *Learn, Practice, Succeed* book puts in students' hands the problem sets and fluency exercises they need to express and consolidate their classroom learning and master grade-level mathematics. Once students learn and practice, they know they can succeed.

What is in the Learn, Practice, Succeed *book?*

Fluency Practice: Our printed fluency activities utilize the format we call a Sprint. Instead of rote recall, Sprints use patterns across a sequence of problems to engage students in reasoning and to reinforce number sense while building speed and accuracy. Sprints are inherently differentiated, with problems building from simple to complex. The tempo of the Sprint provides a low-stakes adrenaline boost that increases memory and automaticity.

Classwork: A carefully sequenced set of examples, exercises, and reflection questions support students' in-class experiences and dialogue. Having classwork preprinted makes efficient use of class time and provides a written record that students can refer to later.

Exit Tickets: Students show teachers what they know through their work on the daily Exit Ticket. This check for understanding provides teachers with valuable real-time evidence of the efficacy of that day's instruction, giving critical insight into where to focus next.

Homework Helpers and Problem Sets: The daily Problem Set gives students additional and varied practice and can be used as differentiated practice or homework. A set of worked examples, Homework Helpers, support students' work on the Problem Set by illustrating the modeling and reasoning the curriculum uses to build understanding of the concepts the lesson addresses.

Homework Helpers and Problem Sets from prior grades or modules can be leveraged to build foundational skills. When coupled with *Affirm®, Eureka Math's* digital assessment system, these Problem Sets enable educators to give targeted practice and to assess student progress. Alignment with the mathematical models and language used across *Eureka Math* ensures that students notice the connections and relevance to their daily instruction, whether they are working on foundational skills or getting extra practice on the current topic.

Where can I learn more about Eureka Math *resources?*

The Great Minds® team is committed to supporting students, families, and educators with an ever-growing library of resources, available at eureka-math.org. The website also offers inspiring stories of success in the *Eureka Math* community. Share your insights and accomplishments with fellow users by becoming a *Eureka Math* Champion.

Best wishes for a year filled with "aha" moments!

Jill Diniz

Jill Diniz
Chief Academic Officer, Mathematics
Great Minds

Contents

Module 4: Expressions and Equations

Topic A: Relationships of the Operations

Lesson 1 1

Lesson 2 13

Lesson 3 25

Lesson 4 35

Topic B: Special Notations of Operations

Lesson 5 45

Lesson 6 57

Topic C: Replacing Letters and Numbers

Lesson 7 71

Lesson 8 83

Topic D: Expanding, Factoring, and Distributing Expressions

Lesson 9 99

Lesson 10 111

Lesson 11 123

Lesson 12 141

Lesson 13 153

Lesson 14 161

Topic E: Expressing Operations in Algebraic Form

Lesson 15 169

Lesson 16 179

Lesson 17 189

Topic F: Writing and Evaluating Expressions and Formulas

Lesson 18 201

Lesson 19 211

Lesson 20 229

Lesson 21 239

Lesson 22 251

Topic G: Solving Equations

Lesson 23 261

Lesson 24 271

Lesson 25 281

Lesson 26 295

Lesson 27 309

Lesson 28 323

Lesson 29 341

Topic H: Applications of Equations

Lesson 30 353

Lesson 31 369

Lesson 32 379

Lesson 33 391

Lesson 34 401

Opening Exercise

a. Draw a tape diagram to represent the following expression: $5 + 4$.

b. Write an expression for each tape diagram.

   i.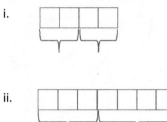

   ii.

Exercises

1. Predict what will happen when a tape diagram has a large number of squares, some squares are removed, and then the same amount of squares are added back on.

2. Build a tape diagram with 10 squares.

    a. Remove six squares. Write an expression to represent the tape diagram.

    b. Add six squares onto the tape diagram. Alter the original expression to represent the current tape diagram.

c. Evaluate the expression.

3. Write an equation, using variables, to represent the identities we demonstrated with tape diagrams.

4. Using your knowledge of identities, fill in each of the blanks.

   a. $4 + 5 - \underline{\hspace{1cm}} = 4$

   b. $25 - \underline{\hspace{1cm}} + 10 = 25$

   c. $\underline{\hspace{1cm}} + 16 - 16 = 45$

   d. $56 - 20 + 20 = \underline{\hspace{1cm}}$

5. Using your knowledge of identities, fill in each of the blanks.

   a. $a + b - \underline{\hspace{1cm}} = a$

   b. $c - d + d = \underline{\hspace{1cm}}$

   c. $e + \underline{\hspace{1cm}} - f = e$

   d. $\underline{\hspace{1cm}} - h + h = g$

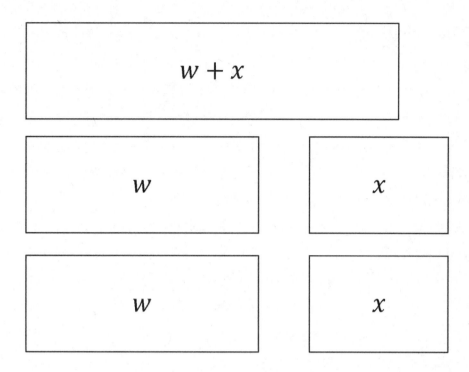

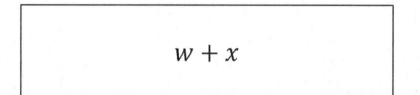

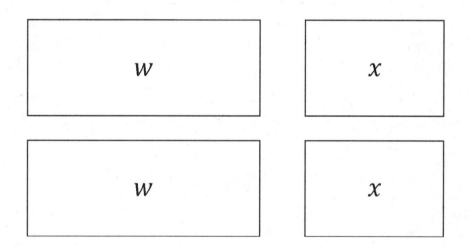

Name _____   Date _____

1. Draw tape diagrams to represent each of the following number sentences.

    a. $3 + 5 - 5 = 3$

    b. $8 - 2 + 2 = 8$

2. Fill in each blank.

    a. $65 + \underline{\hspace{1cm}} - 15 = 65$

    b. $\underline{\hspace{1cm}} + g - g = k$

    c. $a + b - \underline{\hspace{1cm}} = a$

    d. $367 - 93 + 93 = \underline{\hspace{1cm}}$

1. Fill in each blank.

a. _____ + 25 − 25 = 45

   45

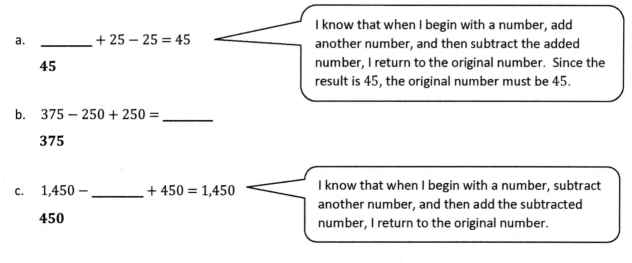

I know that when I begin with a number, add another number, and then subtract the added number, I return to the original number. Since the result is 45, the original number must be 45.

b. 375 − 250 + 250 = _____

   375

c. 1,450 − _____ + 450 = 1,450

   450

I know that when I begin with a number, subtract another number, and then add the subtracted number, I return to the original number.

2. Why are the equations $x + y − y = x$ and $x − y + y = x$ called identities?

   These equations are called identities because the variables can be replaced with any number, and after completing the operations, the result is the original number.

I can test this by replacing the variables with numbers. If I replace x with 10 and y with 4, the resulting number sentence $10 + 4 − 4 = 10$ is true. The other resulting number sentence $10 − 4 + 4 = 10$ is also true.

1. Fill in each blank.

   a. _____ $+ 25 - 25 = 45$

      45

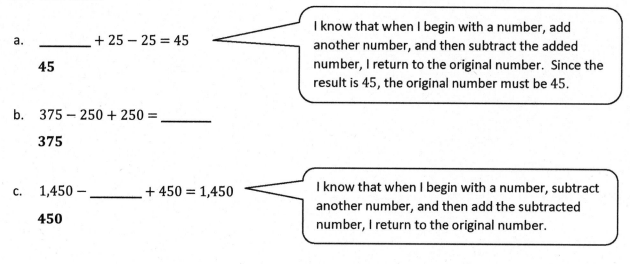

I know that when I begin with a number, add another number, and then subtract the added number, I return to the original number. Since the result is 45, the original number must be 45.

   b. $375 - 250 + 250 =$ _____

      375

   c. $1,450 -$ _____ $+ 450 = 1,450$

      450

I know that when I begin with a number, subtract another number, and then add the subtracted number, I return to the original number.

2. Why are the equations $x + y - y = x$ and $x - y + y = x$ called identities?

   These equations are called identities because the variables can be replaced with any number, and after completing the operations, the result is the original number.

I can test this by replacing the variables with numbers. If I replace x with 10 and y with 4, the resulting number sentence $10 + 4 - 4 = 10$ is true. The other resulting number sentence $10 - 4 + 4 = 10$ is also true.

Lesson 1:   The Relationship of Addition and Subtraction   9

1. Fill in each blank.

   a. _____ $+ 15 - 15 = 21$

   b. $450 - 230 + 230 =$ _____

   c. $1289 -$ _____ $+ 856 = 1289$

2. Why are the equations $w - x + x = w$ and $w + x - x = w$ called *identities*?

Division of Fractions—Round 1

Number Correct: _____

Directions: Evaluate each expression and simplify. Use blank spaces to create like units, where applicable.

1.	9 ones ÷ 3 ones	
2.	$9 \div 3$	
3.	9 tens ÷ 3 tens	
4.	$90 \div 30$	
5.	9 hundreds ÷ 3 hundreds	
6.	$900 \div 300$	
7.	9 halves ÷ 3 halves	
8.	$\dfrac{9}{2} \div \dfrac{3}{2}$	
9.	9 fourths ÷ 3 fourths	
10.	$\dfrac{9}{4} \div \dfrac{3}{4}$	
11.	$\dfrac{9}{8} \div \dfrac{3}{8}$	
12.	$\dfrac{2}{3} \div \dfrac{1}{3}$	
13.	$\dfrac{1}{3} \div \dfrac{2}{3}$	
14.	$\dfrac{6}{7} \div \dfrac{2}{7}$	
15.	$\dfrac{5}{7} \div \dfrac{2}{7}$	
16.	$\dfrac{3}{7} \div \dfrac{4}{7}$	
17.	$\dfrac{6}{10} \div \dfrac{2}{10}$	
18.	$\dfrac{6}{10} \div \dfrac{4}{10}$	
19.	$\dfrac{6}{10} \div \dfrac{8}{10}$	
20.	$\dfrac{7}{12} \div \dfrac{2}{12}$	
21.	$\dfrac{6}{12} \div \dfrac{9}{12}$	
22.	$\dfrac{4}{12} \div \dfrac{11}{12}$	

23.	$\dfrac{6}{10} \div \dfrac{4}{10}$	
24.	$\dfrac{6}{10} \div \dfrac{2}{5} = \dfrac{6}{10} \div \dfrac{\ }{10}$	
25.	$\dfrac{10}{12} \div \dfrac{5}{12}$	
26.	$\dfrac{5}{6} \div \dfrac{5}{12} = \dfrac{\ }{12} \div \dfrac{5}{12}$	
27.	$\dfrac{10}{12} \div \dfrac{3}{12}$	
28.	$\dfrac{10}{12} \div \dfrac{1}{4} = \dfrac{10}{12} \div \dfrac{\ }{12}$	
29.	$\dfrac{5}{6} \div \dfrac{3}{12} = \dfrac{\ }{12} \div \dfrac{3}{12}$	
30.	$\dfrac{5}{10} \div \dfrac{2}{10}$	
31.	$\dfrac{5}{10} \div \dfrac{1}{5} = \dfrac{5}{10} \div \dfrac{\ }{10}$	
32.	$\dfrac{1}{2} \div \dfrac{2}{10} = \dfrac{\ }{10} \div \dfrac{2}{10}$	
33.	$\dfrac{1}{2} \div \dfrac{2}{4}$	
34.	$\dfrac{3}{4} \div \dfrac{2}{8}$	
35.	$\dfrac{1}{2} \div \dfrac{3}{8}$	
36.	$\dfrac{1}{2} \div \dfrac{1}{5} = \dfrac{\ }{10} \div \dfrac{\ }{10}$	
37.	$\dfrac{2}{4} \div \dfrac{1}{3}$	
38.	$\dfrac{1}{4} \div \dfrac{4}{6}$	
39.	$\dfrac{3}{4} \div \dfrac{2}{6}$	
40.	$\dfrac{5}{6} \div \dfrac{1}{4}$	
41.	$\dfrac{2}{9} \div \dfrac{5}{6}$	
42.	$\dfrac{5}{9} \div \dfrac{1}{6}$	
43.	$\dfrac{1}{2} \div \dfrac{1}{7}$	
44.	$\dfrac{5}{7} \div \dfrac{1}{2}$	

Lesson 2: The Relationship of Multiplication and Division

© 2019 Great Minds®. eureka-math.org

Division of Fractions—Round 2

Number Correct: _____

Improvement: _____

Directions: Evaluate each expression and simplify. Use blank spaces to create like units, where applicable.

1.	12 ones ÷ 2 ones	
2.	$12 \div 2$	
3.	12 tens ÷ 2 tens	
4.	$120 \div 20$	
5.	12 hundreds ÷ 2 hundreds	
6.	$1,200 \div 200$	
7.	12 halves ÷ 2 halves	
8.	$\dfrac{12}{2} \div \dfrac{2}{2}$	
9.	12 fourths ÷ 3 fourths	
10.	$\dfrac{12}{4} \div \dfrac{3}{4}$	
11.	$\dfrac{12}{8} \div \dfrac{3}{8}$	
12.	$\dfrac{2}{4} \div \dfrac{1}{4}$	
13.	$\dfrac{1}{4} \div \dfrac{2}{4}$	
14.	$\dfrac{4}{5} \div \dfrac{2}{5}$	
15.	$\dfrac{2}{5} \div \dfrac{4}{5}$	
16.	$\dfrac{3}{5} \div \dfrac{1}{5}$	
17.	$\dfrac{6}{8} \div \dfrac{2}{8}$	
18.	$\dfrac{6}{8} \div \dfrac{4}{8}$	
19.	$\dfrac{6}{8} \div \dfrac{5}{8}$	
20.	$\dfrac{6}{10} \div \dfrac{2}{10}$	
21.	$\dfrac{7}{10} \div \dfrac{8}{10}$	
22.	$\dfrac{4}{10} \div \dfrac{7}{10}$	

23.	$\dfrac{6}{12} \div \dfrac{4}{12}$	
24.	$\dfrac{6}{12} \div \dfrac{2}{6} = \dfrac{6}{12} \div \dfrac{}{12}$	
25.	$\dfrac{8}{14} \div \dfrac{7}{14}$	
26.	$\dfrac{8}{14} \div \dfrac{1}{2} = \dfrac{8}{14} \div \dfrac{}{14}$	
27.	$\dfrac{11}{14} \div \dfrac{2}{14}$	
28.	$\dfrac{11}{14} \div \dfrac{1}{7} = \dfrac{11}{14} \div \dfrac{}{14}$	
29.	$\dfrac{1}{7} \div \dfrac{6}{14} = \dfrac{}{14} \div \dfrac{6}{14}$	
30.	$\dfrac{7}{18} \div \dfrac{3}{18}$	
31.	$\dfrac{7}{18} \div \dfrac{1}{6} = \dfrac{7}{18} \div \dfrac{}{18}$	
32.	$\dfrac{1}{3} \div \dfrac{12}{18} = \dfrac{}{18} \div \dfrac{12}{18}$	
33.	$\dfrac{1}{6} \div \dfrac{4}{18}$	
34.	$\dfrac{4}{12} \div \dfrac{8}{6}$	
35.	$\dfrac{1}{3} \div \dfrac{3}{15}$	
36.	$\dfrac{2}{6} \div \dfrac{1}{9} = \dfrac{}{18} \div \dfrac{}{18}$	
37.	$\dfrac{1}{6} \div \dfrac{4}{9}$	
38.	$\dfrac{2}{3} \div \dfrac{3}{4}$	
39.	$\dfrac{1}{3} \div \dfrac{3}{5}$	
40.	$\dfrac{1}{7} \div \dfrac{1}{2}$	
41.	$\dfrac{5}{6} \div \dfrac{2}{9}$	
42.	$\dfrac{5}{9} \div \dfrac{2}{6}$	
43.	$\dfrac{5}{6} \div \dfrac{4}{9}$	
44.	$\dfrac{1}{2} \div \dfrac{4}{5}$	

Opening Exercise

Draw a pictorial representation of the division and multiplication problems using a tape diagram.

    a. $8 \div 2$

    b. 3×2

Exploratory Challenge

Work in pairs or small groups to determine equations to show the relationship between multiplication and division. Use tape diagrams to provide support for your findings.

1. Create two equations to show the relationship between multiplication and division. These equations should be identities and include variables. Use the squares to develop these equations.

2. Write your equations on large paper. Show a series of tape diagrams to defend each of your equations.

Use the following rubric to critique other posters.

1. Name of the group you are critiquing

2. Equation you are critiquing

3. Whether or not you believe the equations are true and reasons why

Name _____   Date _____

1. Fill in the blanks to make each equation true.

    a. $12 \div 3 \times$ _____ $= 12$

    b. $f \times h \div h =$ _____

    c. $45 \times$ _____ $\div 15 = 45$

    d.   _____ $\div r \times r = p$

2. Draw a series of tape diagrams to represent the following number sentences.
    a. $12 \div 3 \times 3 = 12$

    b. $4 \times 5 \div 5 = 4$

1. Fill in each blank to make each equation true.

   a. $145 \div 5 \times 5 =$ _____

      145

   b. _____ $\div 15 \times 15 = 480$

      480

   If I divide a number by another number and then multiply that result by the number I divided by, my final result is the original number.

   c. $65 \times$ _____ $\div 15 = 65$

      15

   If I multiply a number by another number and then divide that result by the number I multiplied by, my final result is the original number.

   d. $533 \times 13 \div$ _____ $= 533$

      13

2. How is the relationship of multiplication and division similar to the relationship of addition and subtraction?

   Both relationships create identities.

   I can prove this by substituting the variables in the identities with numbers. In the identity $x + y - y = x$, I can replace x with 8 and y with 4. $8 + 4 - 4 = 8$. This is a true equation. This is also true for the relationship between multiplication and division. Using the same replacements in the identity $x \times y \div y = x$, the result is $8 \times 4 \div 4 = 8$, which is a true equation.

1. Fill in each blank to make the equation true.

   a. $132 \div 3 \times 3 = \underline{\qquad}$

   b. $\underline{\qquad} \div 25 \times 25 = 225$

   c. $56 \times \underline{\qquad} \div 8 = 56$

   d. $452 \times 12 \div \underline{\qquad} = 452$

2. How is the relationship of addition and subtraction similar to the relationship of multiplication and division?

Opening Exercise

Write two different expressions that can be depicted by the tape diagram shown. One expression should include addition, while the other should include multiplication.

a.

b.

c.

Exercises

1. Write the addition sentence that describes the model and the multiplication sentence that describes the model.

2. Write an equivalent expression to demonstrate the relationship of multiplication and addition.

   a. $6 + 6$

   b. $3 + 3 + 3 + 3 + 3 + 3$

   c. $4 + 4 + 4 + 4 + 4$

   d. 6×2

   e. 4×6

   f. 3×9

   g. $h + h + h + h + h$

   h. $6y$

Lesson 3:   The Relationship of Multiplication and Addition

3. Roberto is not familiar with tape diagrams and believes that he can show the relationship of multiplication and addition on a number line. Help Roberto demonstrate that the expression 3×2 is equivalent to $2 + 2 + 2$ on a number line.

4. Tell whether the following equations are true or false. Then, explain your reasoning.

   a. $x + 6g - 6g = x$

   b. $2f - 4e + 4e = 2f$

5. Write an equivalent expression to demonstrate the relationship between addition and multiplication.

a. $6 + 6 + 6 + 6 + 4 + 4 + 4$

b. $d + d + d + w + w + w + w + w$

c. $a + a + b + b + b + c + c + c + c$

Name _____   Date _____

Write an equivalent expression to show the relationship of multiplication and addition.

1. $8 + 8 + 8 + 8 + 8 + 8 + 8 + 8 + 8$

2. 4×9

3. $6 + 6 + 6$

4. $7h$

5. $j + j + j + j + j$

6. $u + u + u + u + u + u + u + u + u + u$

Write an equivalent expression to show the relationship between multiplication and addition.

1. $20 + 20 + 20$

    3 × 20

    20 is repeatedly added 3 times.
    This is the same as 3×20.

2. 7×3

    3 + 3 + 3 + 3 + 3 + 3 + 3

    There are 7 copies of 3.
    I can repeatedly add 3 seven times.

3. $8x$

    x + x + x + x + x + x + x + x

    There are 8 copies of x.
    I can repeatedly add x eight times.

4. $f + f + f + f$

    4f

    f is repeatedly added 4 times.
    This is the same as $4 \times f$, or $4f$.

Write an equivalent expression to show the relationship of multiplication and addition.

1. $10 + 10 + 10$

2. $4 + 4 + 4 + 4 + 4 + 4 + 4$

3. 8×2

4. 3×9

5. $6m$

6. $d + d + d + d + d$

Exercise 1

Build subtraction equations using the indicated equations. The first example has been completed for you.

Division Equation	Divisor Indicates the Size of the Unit	Tape Diagram	What is x, y, z?
$12 \div x = 4$	$12 - x - x - x - x = 0$	$12 - 3 - 3 - 3 - 3 = 0; x = 3$ units in each group	$x = 3$
$18 \div x = 3$			
$35 \div y = 5$			
$42 \div z = 6$			

Division Equation	Divisor Indicates the Number of Units	Tape Diagram	What is x, y, z?
$12 \div x = 4$	$12 - 4 - 4 - 4 = 0$	$12 - 4 - 4 - 4 = 0; x = 3$ groups	$x = 3$
$18 \div x = 3$			
$35 \div y = 5$			
$42 \div z = 6$			

Exercise 2

Answer each question using what you have learned about the relationship of division and subtraction.

    a.   If $12 \div x = 3$, how many times would x have to be subtracted from 12 in order for the answer to be zero? What is the value of x?

    b. $36 - f - f - f - f = 0$. Write a division sentence for this repeated subtraction sentence. What is the value of f?

    c.   If $24 \div b = 12$, which number is being subtracted 12 times in order for the answer to be zero?

Graphic Organizer Reproducible

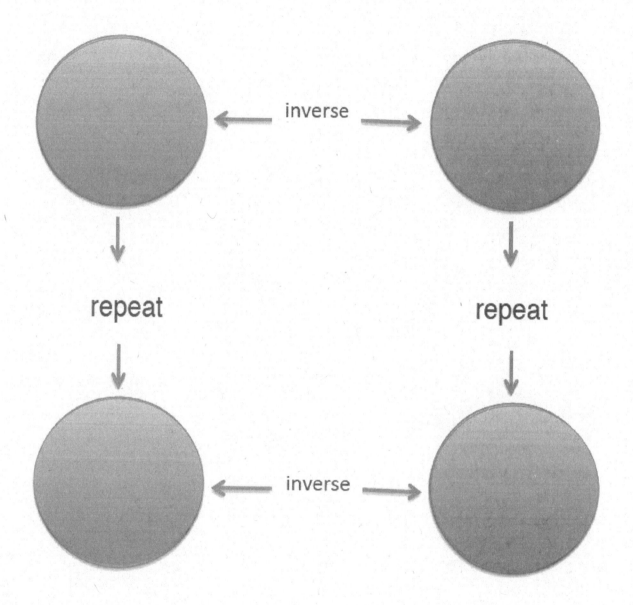

Name _____   Date _____

1. Represent $56 \div 8 = 7$ using subtraction. Explain your reasoning.

2. Explain why $30 \div x = 6$ is the same as $30 - x - x - x - x - x - x = 0$. What is the value of x in this example?

Name _____ Date _____

1. Represent $56 \div 8 = 7$ using subtraction. Explain your reasoning.

2. Explain why $30 \div x = 6$ is the same as $30 - x - x - x - x - x - x = 0$. What is the value of x in this example?

Build subtraction equations using the indicated equations.

> The quotient in each of these equations represents the number of groups.

> The number of groups is represented in the tape diagrams, as well.

	Division Equation	Divisor Indicates the Size of the Unit	Tape Diagram	What is x or y?
1.	$10 \div x = 2$	$10 - x - x = 0$		$x = 5$
2.	$24 \div x = 3$	$24 - x - x - x = 0$		$x = 8$
3.	$35 \div y = 5$	$35 - y - y - y - y - y = 0$		$y = 7$

> The quotient is also represented by the number of times the divisor is repeatedly subtracted from the dividend. The number that is being repeatedly subtracted is the dividend.

> The divisor is the number in each of the groups.

The quotient in each of these equations represents the number that is repeatedly subtracted from the dividend.

The number that is repeatedly being subtracted (the quotient) is represented in the tape diagrams, as well.

	Division Equation	Divisor Indicates the Number of Units	Tape Diagram	What is x or y?
1.	$10 \div x = 2$	$10 - 2 - 2 - 2 - 2 - 2 = 0$		$x = 5$
2.	$24 \div x = 3$	$24 - 3 - 3 - 3 - 3 - 3 - 3 - 3 - 3 = 0$		$x = 8$
3.	$35 \div y = 5$	$35 - 5 - 5 - 5 - 5 - 5 - 5 - 5 = 0$		$y = 7$

The quotient is being repeatedly subtracted from the dividend. The total number of times it was subtracted is the divisor.

The divisor is the number of times, or the number of groups of, the quotient that was repeatedly subtracted from the dividend.

Lesson 4:   The Relationship of Division and Subtraction

Build subtraction equations using the indicated equations.

	Division Equation	Divisor Indicates the Size of the Unit	Tape Diagram	What is x, y, z?
1.	$24 \div x = 4$			
2.	$36 \div x = 6$			
3.	$28 \div y = 7$			
4.	$30 \div y = 5$			
5.	$16 \div z = 4$			

	Division Equation	Divisor Indicates the Number of Units	Tape Diagram	What is x, y, z?
1.	$24 \div x = 4$			
2.	$36 \div x = 6$			
3.	$28 \div y = 7$			
4.	$30 \div y = 5$			
5.	$16 \div z = 4$			

Opening Exercise

As you evaluate these expressions, pay attention to how you arrived at your answers.

$4 + 4 + 4 + 4 + 4 + 4 + 4 + 4 + 4 + 4$

$9 + 9 + 9 + 9 + 9$

$10 + 10 + 10 + 10 + 10$

Examples 1–10

Write each expression in exponential form.

1. $5 \times 5 \times 5 \times 5 \times 5 =$

2. $2 \times 2 \times 2 \times 2 =$

Write each expression in expanded form.

3. $8^3 =$

4. $10^6 =$

5. $g^3 =$

Go back to Examples 1–4, and use a calculator to evaluate the expressions.

What is the difference between $3g$ and g^3?

6. Write the expression in expanded form, and then evaluate.

    $(3.8)^4 =$

7. Write the expression in exponential form, and then evaluate.

    $2.1 \times 2.1 =$

8. Write the expression in exponential form, and then evaluate.

    $0.75 \times 0.75 \times 0.75 =$

The base number can also be a fraction. Convert the decimals to fractions in Examples 7 and 8 and evaluate. Leave your answer as a fraction. Remember how to multiply fractions!

9. Write the expression in exponential form, and then evaluate.

$$\frac{1}{2} \times \frac{1}{2} \times \frac{1}{2} =$$

10. Write the expression in expanded form, and then evaluate.

$$\left(\frac{2}{3}\right)^2 =$$

Exercises

1. Fill in the missing expressions for each row. For whole number and decimal bases, use a calculator to find the standard form of the number. For fraction bases, leave your answer as a fraction.

Exponential Form	Expanded Form	Standard Form
3^2	3×3	9
	$2 \times 2 \times 2 \times 2 \times 2 \times 2$	
4^5		
	$\frac{3}{4} \times \frac{3}{4}$	
	1.5×1.5	

2. Write five cubed in all three forms: exponential form, expanded form, and standard form.

3. Write fourteen and seven-tenths squared in all three forms.

4. One student thought two to the third power was equal to six. What mistake do you think he made, and how would you help him fix his mistake?

Lesson 5:   Exponents

Lesson Summary

EXPONENTIAL NOTATION FOR WHOLE NUMBER EXPONENTS: Let m be a nonzero whole number. For any number a, the expression a^m is the product of m factors of a, i.e.,

$$a^m = \underbrace{a \cdot a \cdots \cdot a}_{m \text{ times}}.$$

The number a is called the *base*, and m is called the *exponent* or *power* of a.

When m is 1, "the product of one factor of a" just means a (i.e., $a^1 = a$). Raising any nonzero number a to the power of 0 is defined to be 1 (i.e., $a^0 = 1$ for all $a \neq 0$).

Name _____ Date _____

1. What is the difference between $6z$ and z^6?

2. Write 10^3 as a multiplication expression having repeated factors.

3. Write $8 \times 8 \times 8 \times 8$ using an exponent.

Name _____   Date _____

1. What is the difference between $6z$ and z^6?

2. Write 10^3 as a multiplication expression having repeated factors.

3. Write $8 \times 8 \times 8 \times 8$ using an exponent.

1. Complete the table by filling in the blank cells. Use a calculator when needed.

Exponential Form	Expanded Form	Standard Form
2^3	$2 \times 2 \times 2$	8
5^4	$5 \times 5 \times 5 \times 5$	625
$(1.5)^2$	1.5×1.5	2.25
$\left(\dfrac{1}{3}\right)^5$	$\dfrac{1}{3} \times \dfrac{1}{3} \times \dfrac{1}{3} \times \dfrac{1}{3} \times \dfrac{1}{3}$	$\dfrac{1}{243}$

> When I am given exponential form, I can expand by multiplying the base factor by itself the number of times the exponent states. Then, I can evaluate the multiplication expression. When I am given the expanded form, I note the factor being multiplied as the base and then count the number of times it is being multiplied. That number represents the exponent.

2. Why do whole numbers raised to an exponent get greater, while fractions raised to an exponent get smaller?

   As whole numbers are multiplied by themselves, products are larger because there are more groups. As fractions of fractions are taken, the product is smaller. A part of a part is less than how much was started with.

3. The powers of 3 that are in the range 3 through 1,000 are 3, 9, 27, 81, 243, and 729. Find all the powers of 4 that are in the range 4 through 1,000.

   4, 16, 64, 256

> I begin with the base factor and continue to multiply it by itself repeatedly until I determine the last product before I reach 1,000.

4. Find all the powers of 5 in the range 5 through 1,000.

   5, 25, 125, 625

5. Write an equivalent expression for $x \times y$ using only addition.

   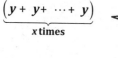

> Because multiplication is repeated addition, I add y to itself the number of times x states.

6. Write an equivalent expression for n^t using only multiplication.

$$\underbrace{(n \cdot n \cdot \ldots \cdot n)}_{t \text{ times}}$$

> Because a number to a power is repeated multiplication, I multiply the base factor n by itself the number of times t states.

   a. Explain what n is in this new expression.

   n is the factor that is repeatedly multiplied by itself.

   b. Explain what t is in this new expression.

   t is the number of times n will be multiplied.

7. What is the advantage of using exponential notation?

   It is a more efficient way of writing a multiplication expression if the factors are all the same.

8. What is the difference between $5x$ and x^5? Evaluate both of these expressions when $x = 3$.

   $5x$ means five times x. This is the same as $x + x + x + x + x$. x^5 means x to the fifth power, or $x \cdot x \cdot x \cdot x \cdot x$.

   When $x = 3$, $5x = 5 \cdot 3 = 15$.

   When $x = 3$, $x^5 = 3 \cdot 3 \cdot 3 \cdot 3 \cdot 3 = 243$.

EUREKA MATH

1. Complete the table by filling in the blank cells. Use a calculator when needed.

Exponential Form	Expanded Form	Standard Form
3^5		
	$4 \times 4 \times 4$	
$(1.9)^2$		
$\left(\dfrac{1}{2}\right)^5$		

2. Why do whole numbers raised to an exponent get greater, while fractions raised to an exponent get smaller?

3. The powers of 2 that are in the range 2 through 1,000 are 2, 4, 8, 16, 32, 64, 128, 256, and 512. Find all the powers of 3 that are in the range 3 through 1,000.

4. Find all the powers of 4 in the range 4 through 1,000.

5. Write an equivalent expression for $n \times a$ using only addition.

6. Write an equivalent expression for w^b using only multiplication.
   a. Explain what w is in this new expression.
   b. Explain what b is in this new expression.

7. What is the advantage of using exponential notation?

8. What is the difference between $4x$ and x^4? Evaluate both of these expressions when $x = 2$.

Example 1: Expressions with Only Addition, Subtraction, Multiplication, and Division

What operations are evaluated first?

What operations are always evaluated last?

Exercises 1–3

1. $4 + 2 \times 7$

2. $36 \div 3 \times 4$

3. $20 - 5 \times 2$

Example 2: Expressions with Four Operations and Exponents

$$4 + 9^2 \div 3 \times 2 - 2$$

What operation is evaluated first?

What operations are evaluated next?

What operations are always evaluated last?

What is the final answer?

Exercises 4–5

4. $90 - 5^2 \times 3$

5. $4^3 + 2 \times 8$

Lesson 6:   The Order of Operations

Example 3: Expressions with Parentheses

Consider a family of 4 that goes to a soccer game. Tickets are $5.00 each. The mom also buys a soft drink for $2.00. How would you write this expression?

How much will this outing cost?

Consider a different scenario: The same family goes to the game as before, but each of the family members wants a drink. How would you write this expression?

Why would you add the 5 and 2 first?

How much will this outing cost?

How many groups are there?

What does each group comprise?

Exercises 6–7

6. $2 + (9^2 - 4)$

7. $2 \cdot \left(13 + 5 - 14 \div (3 + 4)\right)$

$$2 \times (3 + 4^2)$$

Which value will we evaluate first within the parentheses? Evaluate.

Evaluate the rest of the expression.

What do you think will happen when the exponent in this expression is outside of the parentheses?

$$2 \times (3 + 4)^2$$

Will the answer be the same?

Which should we evaluate first? Evaluate.

What happens differently here than in our last example?

What should our next step be?

Evaluate to find the final answer.

What do you notice about the two answers?

What was different between the two expressions?

What conclusions can you draw about evaluating expressions with parentheses and exponents?

Exercises 8–9

8. $7 + (12 - 3^2)$

9. $7 + (12 - 3)^2$

Lesson 6:   The Order of Operations

Lesson Summary

NUMERICAL EXPRESSION: A *numerical expression* is a number, or it is any combination of sums, differences, products, or divisions of numbers that evaluates to a number.

Statements like "3 +" or "3 ÷ 0" are not numerical expressions because neither represents a point on the number line. Note: Raising numbers to whole number powers are considered numerical expressions as well since the operation is just an abbreviated form of multiplication, e.g., $2^3 = 2 \cdot 2 \cdot 2$.

VALUE OF A NUMERICAL EXPRESSION: The *value of a numerical expression* is the number found by evaluating the expression.

For example: $\frac{1}{3} \cdot (2 + 4) + 7$ is a numerical expression, and its value is 9.

Name _____   Date _____

1. Evaluate this expression: $39 \div (2 + 1) - 2 \times (4 + 1)$.

2. Evaluate this expression: $12 \times (3 + 2^2) \div 2 - 10$.

3. Evaluate this expression: $12 \times (3 + 2)^2 \div 2 - 10$.

Evaluate each expression.

1. $2 \times 4 + 1 \times 7 + 1$

    $8 + 7 + 1$

    16

> I know that multiplication is repeated addition and should be evaluated first in this problem. Then I can find the sum of the resulting addition expression.

2. $(\$1.50 + 2 \times \$0.75 + 5 \times \$0.01) \times 20$

    $(\$1.50 + \$1.50 + \$0.05) \times 20$

    $\$3.05 \times 20$

    $\$61$

> I need to evaluate the expressions within the parentheses first. The most powerful operation in the parentheses is multiplication. I will multiply first and then have a resulting addition expression within the parentheses. From there I will evaluate the addition expression in the parentheses first and then multiply by 20.

3. $(3 \times 7) + (7 \times 2) + 2$

    $21 + 14 + 2$

    37

> I know sometimes parentheses group parts of an expression for clarity. In this problem, the parentheses are actually not necessary since the operation of multiplication would be evaluated first.

4. $\left((15 \div 5)^2 - (27 \div 3^2)\right) \times (6 \div 3)$

    $\left((3)^2 - (27 \div 9)\right) \times 2$

    $(9 - 3) \times 2$

    6×2

    12

> I know that I have to evaluate what is in the parentheses first. But in this problem, exponents are in different places—outside parentheses and inside parentheses. I need to evaluate the exponent inside the parentheses before I can evaluate the expressions inside the parentheses.

Evaluate each expression.

1. $3 \times 5 + 2 \times 8 + 2$

2. $(\$1.75 + 2 \times \$0.25 + 5 \times \$0.05) \times 24$

3. $(2 \times 6) + (8 \times 4) + 1$

4. $\big((8 \times 1.95) + (3 \times 2.95) + 10.95\big) \times 1.06$

5. $\big((12 \div 3)^2 - (18 \div 3^2)\big) \times (4 \div 2)$

Example 1

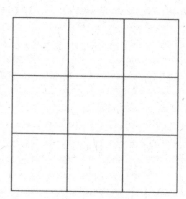

What is the length of one side of this square?

What is the formula for the area of a square?

What is the square's area as a multiplication expression?

What is the square's area?

We can count the units. However, look at this other square. Its side length is 23 cm. That is just too many tiny units to draw. What expression can we build to find this square's area?

What is the area of the square? Use a calculator if you need to.

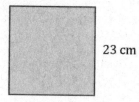

23 cm

Exercise 1

Complete the table below for both squares. Note: These drawings are not to scale.

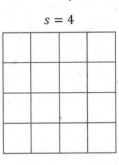

$s = 4$

$s = 25$ in.

Length of One Side of the Square	Square's Area Written as an Expression	Square's Area Written as a Number

Example 2

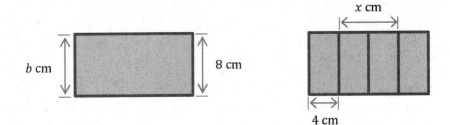

b cm

8 cm

x cm

4 cm

What does the letter b represent in this blue rectangle?

With a partner, answer the following question: Given that the second rectangle is divided into four *equal* parts, what number does the x represent?

How did you arrive at this answer?

What is the total length of the second rectangle? Tell a partner how you know.

If the two large rectangles have equal lengths and widths, find the area of each rectangle.

Discuss with your partner how the formulas for the area of squares and rectangles can be used to evaluate area for a particular figure.

Exercise 2

Complete the table below for both rectangles. Note: These drawings are not to scale. Using a calculator is appropriate.

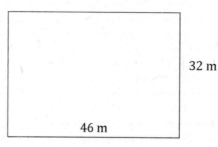

32 m

46 m

Length of Rectangle	Width of Rectangle	Rectangle's Area Written as an Expression	Rectangle's Area Written as a Number

Example 3

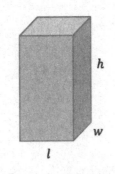

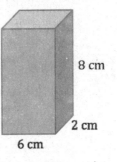

What does the *l* represent in the first diagram?

What does the *w* represent in the first diagram?

What does the *h* represent in the first diagram?

Since we know the formula to find the volume is $V = l \times w \times h$, what number can we substitute for the *l* in the formula? Why?

What other number can we substitute for the *l*?

What number can we substitute for the *w* in the formula? Why?

Lesson 7:   Replacing Letters with Numbers

What number can we substitute for the h in the formula?

Determine the volume of the second right rectangular prism by replacing the letters in the formula with their appropriate numbers.

Exercise 3

Complete the table for both figures. Using a calculator is appropriate.

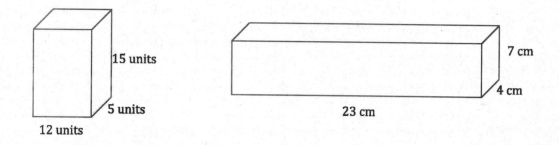

Length of Rectangular Prism	Width of Rectangular Prism	Height of Rectangular Prism	Rectangular Prism's Volume Written as an Expression	Rectangular Prism's Volume Written as a Number

---

Lesson Summary

VARIABLE (description): A *variable* is a symbol (such as a letter) that is a placeholder for a number.

EXPRESSION (description): An *expression* is a numerical expression, or it is the result of replacing some (or all) of the numbers in a numerical expression with variables.

There are two ways to build expressions:

1. We can start out with a numerical expression, like $\frac{1}{3} \cdot (2 + 4) + 7$, and replace some of the numbers with letters to get $\frac{1}{3} \cdot (x + y) + z$.

2. We can build such expressions from scratch, as in $x + x(y - z)$, and note that if numbers were placed in the expression for the variables x, y, and z, the result would be a numerical expression.

Lesson 7:   Replacing Letters with Numbers

Name _____ Date _____

1. In the drawing below, what do the letters l and w represent?

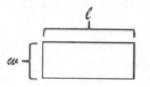

2. What does the expression $l + w + l + w$ represent?

3. What does the expression $l \cdot w$ represent?

4. The rectangle below is congruent to the rectangle shown in Problem 1. Use this information to evaluate the expressions from Problems 2 and 3.

1. Replace the side length of this square with 7 in., and find the area.

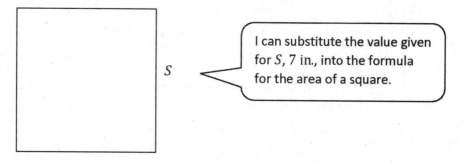

If the side length of the square is 7 in., the area of the square, S^2, is $(7 \text{ in.})^2$ or $7 \text{ in.} \times 7 \text{ in.} = 49 \text{ in}^2$.

2. Complete the table for each of the given figures.

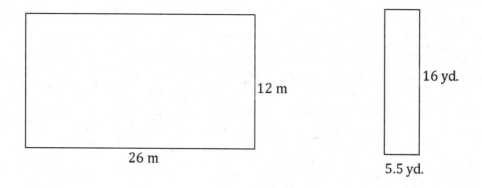

Length of Rectangle	Width of Rectangle	Rectangle's Area Written as an Expression	Rectangle's Area Written as a Number
26 m	12 m	26 m × 12 m	312 m²
16 yd.	5.5 yd.	16 yd. × 5.5 yd.	88 yd²

3. Find the perimeter of each quadrilateral in Problems 1 and 2.

   Problem 1: $P = 28$ **in.**

   Problem 2: $P = 76$ **m;** $P = 43$ **yd.**

   > I can use the formula for perimeter $(l + w + l + w)$, substitute the values for the length and width of each rectangle, and then add.

4. Using the formula $V = l \times w \times h$, find the volume of a right rectangular prism when the length of the prism is 38 cm, the width is 10 cm, and the height is 6 cm.

   $V = l \times w \times h; V = 38 \textbf{ cm} \times 10 \textbf{ cm} \times 6 \textbf{ cm} = 2{,}280 \textbf{ cm}^3$

   > Using the formula, I can substitute the values given in the problem for length (l), width (w), and height (h). When I multiply the numbers, I can use the commutative property to rearrange the order of the numbers, multiply 38×6, and then multiply the result by 10.

1. Replace the side length of this square with 4 in., and find the area.

2. Complete the table for each of the given figures.

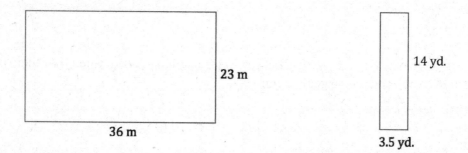

Length of Rectangle	Width of Rectangle	Rectangle's Area Written as an Expression	Rectangle's Area Written as a Number

3. Find the perimeter of each quadrilateral in Problems 1 and 2.

4. Using the formula $V = l \times w \times h$, find the volume of a right rectangular prism when the length of the prism is 45 cm, the width is 12 cm, and the height is 10 cm.

Number Correct: _____

Division of Fractions II—Round 1

Directions: Determine the quotient of the fractions and simplify.

1.	$\dfrac{4}{10} \div \dfrac{2}{10}$	
2.	$\dfrac{9}{12} \div \dfrac{3}{12}$	
3.	$\dfrac{6}{10} \div \dfrac{4}{10}$	
4.	$\dfrac{2}{8} \div \dfrac{3}{8}$	
5.	$\dfrac{2}{7} \div \dfrac{6}{7}$	
6.	$\dfrac{11}{9} \div \dfrac{8}{9}$	
7.	$\dfrac{5}{13} \div \dfrac{10}{13}$	
8.	$\dfrac{7}{8} \div \dfrac{13}{16}$	
9.	$\dfrac{3}{5} \div \dfrac{7}{10}$	
10.	$\dfrac{9}{30} \div \dfrac{3}{5}$	
11.	$\dfrac{1}{3} \div \dfrac{4}{5}$	
12.	$\dfrac{2}{5} \div \dfrac{3}{4}$	
13.	$\dfrac{3}{4} \div \dfrac{5}{9}$	
14.	$\dfrac{4}{5} \div \dfrac{7}{12}$	
15.	$\dfrac{3}{8} \div \dfrac{5}{2}$	

16.	$3\dfrac{1}{8} \div \dfrac{2}{3}$	
17.	$1\dfrac{5}{6} \div \dfrac{1}{2}$	
18.	$\dfrac{5}{8} \div 2\dfrac{3}{4}$	
19.	$\dfrac{1}{3} \div 1\dfrac{4}{5}$	
20.	$\dfrac{3}{4} \div 2\dfrac{3}{10}$	
21.	$2\dfrac{1}{5} \div 1\dfrac{1}{6}$	
22.	$2\dfrac{4}{9} \div 1\dfrac{3}{5}$	
23.	$1\dfrac{2}{9} \div 3\dfrac{2}{5}$	
24.	$2\dfrac{2}{3} \div 3$	
25.	$1\dfrac{3}{4} \div 2\dfrac{2}{5}$	
26.	$4 \div 1\dfrac{2}{9}$	
27.	$3\dfrac{1}{5} \div 6$	
28.	$2\dfrac{5}{6} \div 1\dfrac{1}{3}$	
29.	$10\dfrac{2}{3} \div 8$	
30.	$15 \div 2\dfrac{3}{5}$	

Number Correct: _____

Improvement: _____

Division of Fractions II—Round 2

Directions: Determine the quotient of the fractions and simplify.

1.	$\dfrac{10}{2} \div \dfrac{5}{2}$	
2.	$\dfrac{6}{5} \div \dfrac{3}{5}$	
3.	$\dfrac{10}{7} \div \dfrac{2}{7}$	
4.	$\dfrac{3}{8} \div \dfrac{5}{8}$	
5.	$\dfrac{1}{4} \div \dfrac{3}{12}$	
6.	$\dfrac{7}{5} \div \dfrac{3}{10}$	
7.	$\dfrac{8}{15} \div \dfrac{4}{5}$	
8.	$\dfrac{5}{6} \div \dfrac{5}{12}$	
9.	$\dfrac{3}{5} \div \dfrac{7}{9}$	
10.	$\dfrac{3}{10} \div \dfrac{3}{9}$	
11.	$\dfrac{3}{4} \div \dfrac{7}{9}$	
12.	$\dfrac{7}{10} \div \dfrac{3}{8}$	
13.	$4 \div \dfrac{4}{9}$	
14.	$\dfrac{5}{8} \div 7$	
15.	$9 \div \dfrac{2}{3}$	

16.	$\dfrac{5}{8} \div 1\dfrac{3}{4}$	
17.	$\dfrac{1}{4} \div 2\dfrac{2}{5}$	
18.	$2\dfrac{3}{5} \div \dfrac{3}{8}$	
19.	$1\dfrac{3}{5} \div \dfrac{2}{9}$	
20.	$4 \div 2\dfrac{3}{8}$	
21.	$1\dfrac{1}{2} \div 5$	
22.	$3\dfrac{1}{3} \div 1\dfrac{3}{4}$	
23.	$2\dfrac{2}{5} \div 1\dfrac{1}{4}$	
24.	$3\dfrac{1}{2} \div 2\dfrac{2}{3}$	
25.	$1\dfrac{4}{5} \div 2\dfrac{3}{4}$	
26.	$3\dfrac{1}{6} \div 1\dfrac{3}{5}$	
27.	$3\dfrac{3}{5} \div 2\dfrac{1}{8}$	
28.	$5 \div 1\dfrac{1}{6}$	
29.	$3\dfrac{3}{4} \div 5\dfrac{1}{2}$	
30.	$4\dfrac{2}{3} \div 5\dfrac{1}{4}$	

Opening Exercise

$$4 + 0 = 4$$
$$4 \times 1 = 4$$
$$4 \div 1 = 4$$
$$4 \times 0 = 0$$
$$1 \div 4 = \frac{1}{4}$$

How many of these statements are true?

How many of those statements would be true if the number 4 was replaced with the number 7 in each of the number sentences?

Would the number sentences be true if we were to replace the number 4 with any other number?

What if we replaced the number 4 with the number 0? Would each of the number sentences be true?

What if we replace the number 4 with a letter g? Please write all 4 expressions below, replacing each 4 with a g.

Are these all true (except for $g = 0$) when dividing?

Example 1: Additive Identity Property of Zero

$$g + 0 = g$$

Remember a letter in a mathematical expression represents a number. Can we replace g with any number?

Choose a value for g, and replace g with that number in the equation. What do you observe?

Repeat this process several times, each time choosing a different number for g.

Will all values of g result in a true number sentence?

Write the mathematical language for this property below:

Example 2: Multiplicative Identity Property of One

$$g \times 1 = g$$

Remember a letter in a mathematical expression represents a number. Can we replace g with any number?

Choose a value for g, and replace g with that number in the equation. What do you observe?

Will all values of g result in a true number sentence? Experiment with different values before making your claim.

Write the mathematical language for this property below:

Example 3: Commutative Property of Addition and Multiplication

$$3 + 4 = 4 + 3$$
$$3 \times 4 = 4 \times 3$$

Replace the 3's in these number sentences with the letter a.

Choose a value for a, and replace a with that number in each of the equations. What do you observe?

Lesson 8:   Replacing Numbers with Letters   **89**

© 2019 Great Minds®. eureka-math.org

Will all values of a result in a true number sentence? Experiment with different values before making your claim.

Now, write the equations again, this time replacing the number 4 with a variable, b.

Will all values of a and b result in true number sentences for the first two equations? Experiment with different values before making your claim.

Write the mathematical language for this property below:

Example 4

$$3 + 3 + 3 + 3 = 4 \times 3$$
$$3 \div 4 = \frac{3}{4}$$

Replace the 3's in these number sentences with the letter a.

Lesson 8:   Replacing Numbers with Letters

Choose a value for a, and replace a with that number in each of the equations. What do you observe?

Will all values of a result in a true number sentence? Experiment with different values before making your claim.

Now, write the equations again, this time replacing the number 4 with a variable, b.

Will all values of a and b result in true number sentences for the equations? Experiment with different values before making your claim.

Name _____ Date _____

1. State the commutative property of addition, and provide an example using two different numbers.

2. State the commutative property of multiplication, and provide an example using two different numbers.

3. State the additive property of zero, and provide an example using any other number.

4. State the multiplicative identity property of one, and provide an example using any other number.

1. Demonstrate the property listed in the first column by filling in the third column of the table.

$a \times b = b \times a$

Commutative Property of Addition	$37 + c =$	$c + 37$	$a + b = b + a$
Commutative Property of Multiplication	$m \times n =$	$n \times m$	
Additive Property of Zero	$p + 0 =$	p	$b + 0 = b$
Multiplicative Identity Property of One	$t \times 1 =$	t	

$b \times 1 = b$

2. Why is there no commutative property for subtraction or division? Show examples.

Here is an example of why the commutative property does not work for division. $12 \div 4$ and $4 \div 12$. $12 \div 4 = 3$, but $4 \div 12 = \frac{1}{3}$. For subtraction, the order is important because it can result in different answers. $9 - 2 = 7$, but $2 - 9 = -7$.

1. State the commutative property of addition using the variables a and b.

2. State the commutative property of multiplication using the variables a and b.

3. State the additive property of zero using the variable b.

4. State the multiplicative identity property of one using the variable b.

5. Demonstrate the property listed in the first column by filling in the third column of the table.

Commutative Property of Addition	$25 + c =$	
Commutative Property of Multiplication	$l \times w =$	
Additive Property of Zero	$h + 0 =$	
Multiplicative Identity Property of One	$v \times 1 =$	

6. Why is there no commutative property for subtraction or division? Show examples.

Example 1

Create a bar diagram to show 3 plus 5.

How would this look if you were asked to show 5 plus 3?

Are these two expressions equivalent?

Example 2

How can we show a number increased by 2?

Can you prove this using a model? If so, draw the model.

Example 3

Write an expression to show the sum of m and k.

Which property can be used in Examples 1–3 to show that both expressions given are equivalent?

Example 4

How can we show 10 minus 6?

- Draw a bar diagram to model this expression.

- What expression would represent this model?

- Could we also use $6 - 10$?

Example 5

How can we write an expression to show 3 less than a number?

- Start by drawing a diagram to model the subtraction. Are we taking away from the 3 or the unknown number?

- What expression would represent this model?

Example 6

How would we write an expression to show the number c being subtracted from the sum of a and b?

- Start by writing an expression for "the sum of a and b."

- Now, show c being subtracted from the sum.

Example 7

Write an expression to show the number c minus the sum of a and b.

Why are the parentheses necessary in this example and not the others?

Replace the variables with numbers to see if $c - (a + b)$ is the same as $c - a + b$.

Exercises

1. Write an expression to show the sum of 7 and 1.5.

2. Write two expressions to show w increased by 4. Then, draw models to prove that both expressions represent the same thing.

3. Write an expression to show the sum of a, b, and c.

4. Write an expression and a model showing 3 less than p.

5. Write an expression to show the difference of 3 and p.

6. Write an expression to show 4 less than the sum of g and 5.

7. Write an expression to show 4 decreased by the sum of g and 5.

8. Should Exercises 6 and 7 have different expressions? Why or why not?

Name _____   Date _____

1. Write an expression showing the sum of 8 and a number f.

2. Write an expression showing 5 less than the number k.

3. Write an expression showing the sum of a number h and a number w minus 11.

1. Write two expressions to show a number increased by 6. Then, draw models to prove that both expressions represent the same thing.

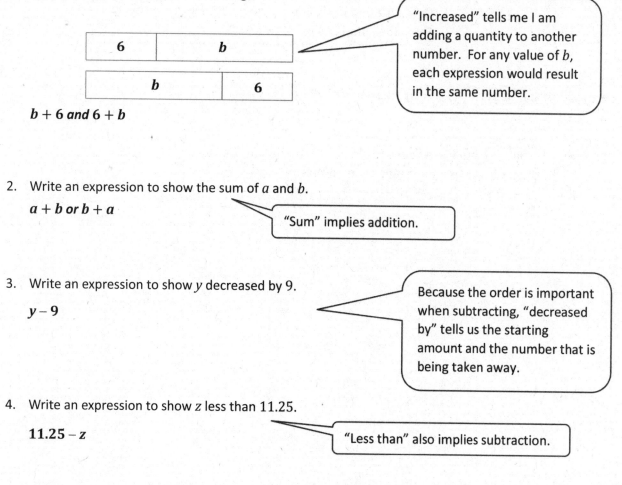

"Increased" tells me I am adding a quantity to another number. For any value of b, each expression would result in the same number.

$b + 6$ *and* $6 + b$

2. Write an expression to show the sum of a and b.

$a + b$ *or* $b + a$

"Sum" implies addition.

3. Write an expression to show y decreased by 9.

$y - 9$

Because the order is important when subtracting, "decreased by" tells us the starting amount and the number that is being taken away.

4. Write an expression to show z less than 11.25.

$11.25 - z$

"Less than" also implies subtraction.

5. Write an expression to show the sum of r and m reduced by 21.

$r + m - 21$

Writing this expression requires two steps. First, r and m are being added. Then, the sum is the starting amount in a subtraction problem.

6. Write an expression to show 4 less than l, plus e.

$l + 4 + e$

> First, l is the starting amount, and 4 is being taken away. Then, the difference is added to e.

7. Write an expression to show 3 less than the sum of p and n.

$p + n - 3$

> First, p and n are added together. Then, the sum is the starting amount in a subtraction problem.

1. Write two expressions to show a number increased by 11. Then, draw models to prove that both expressions represent the same thing.

2. Write an expression to show the sum of x and y.

3. Write an expression to show h decreased by 13.

4. Write an expression to show k less than 3.5.

5. Write an expression to show the sum of g and h reduced by 11.

6. Write an expression to show 5 less than y, plus g.

7. Write an expression to show 5 less than the sum of y and g.

Example 1

Write each expression using the fewest number of symbols and characters. Use math terms to describe the expressions and parts of the expressions.

   a. $6 \times b$

   b. $4 \cdot 3 \cdot h$

   c. $2 \times 2 \times 2 \times a \times b$

   d. $5 \times m \times 3 \times p$

   e. $1 \times g \times w$

Example 2

To expand multiplication expressions, we will rewrite the expressions by including the " · " back into the expressions.

   a. $5g$

   b. $7abc$

   c. $12g$

   d. $3h \cdot 8$

   e. $7g \cdot 9h$

Example 3

   a. Find the product of $4f \cdot 7g$.

   b. Multiply $3de \cdot 9yz$.

   c. Double the product of $6y$ and $3bc$.

      Lesson 10:   Writing and Expanding Multiplication Expressions

---

Lesson Summary

AN EXPRESSION IN EXPANDED FORM: An expression that is written as sums (and/or differences) of products whose factors are numbers, variables, or variables raised to whole number powers is said to be in *expanded form*. A single number, variable, or a single product of numbers and/or variables is also considered to be in expanded form.

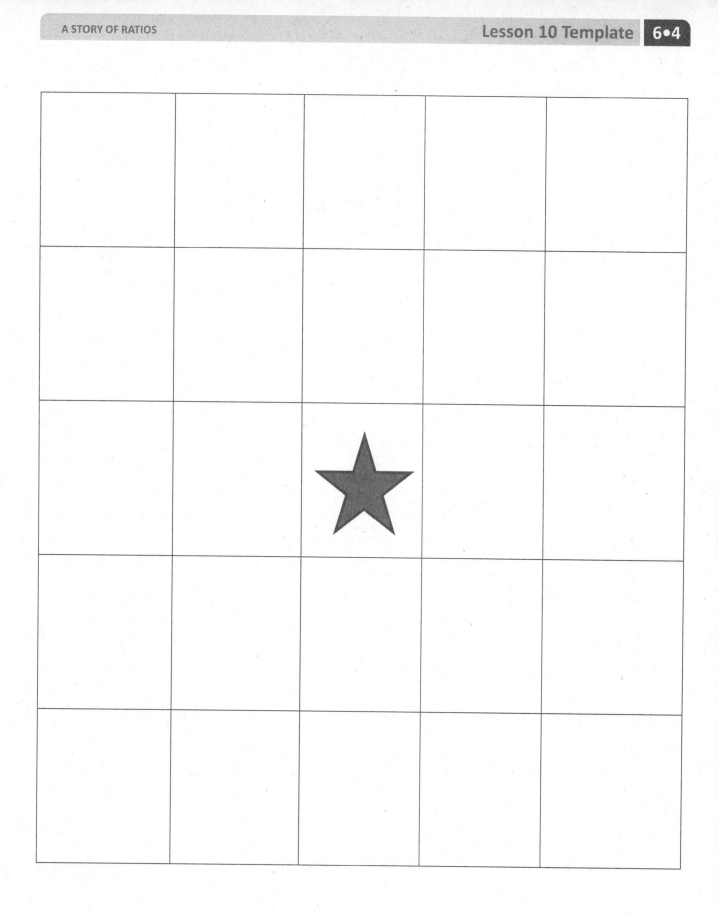

Name _____ Date _____

1. Rewrite the expression in standard form (use the fewest number of symbols and characters possible).

   a. $5g \cdot 7h$

   b. $3 \cdot 4 \cdot 5 \cdot m \cdot n$

2. Name the parts of the expression. Then, write it in expanded form.
   a. $14b$

   b. $30jk$

1. Rewrite the expression in standard form. Use the fewest number of symbols and characters possible.

    a. $6 \cdot 3 \cdot a \cdot b$

      18ab

    b. $4 \cdot 5 \cdot 2 \cdot 10 \cdot y$

      400y

      > When I write an expression in standard form, I do not use the operation symbol or symbols for multiplication. I write the factors next to each other. When possible, I multiply numbers together before writing the product next to the variable or variables.

2. Write the following expressions in expanded form.

    a. $26yz$

      $26 \cdot y \cdot z$ or $2 \cdot 13 \cdot y \cdot z$

      > When I write an expression in expanded form, I write the expression as a product of the factors using the " $\cdot$ " symbol for multiplication.

    b. $12xyz$

      $12 \cdot x \cdot y \cdot z$ or $2 \cdot 2 \cdot 3 \cdot x \cdot y \cdot z$

3. Find the product.

    a. $9a \cdot 3b$

      $27ab$

      > I multiply the coefficients and then write the rest of the variables in alphabetical order.

    b. $6ab \cdot 11cd$

      $66abcd$

1. Rewrite the expression in standard form (use the fewest number of symbols and characters possible).

   a. $5 \cdot y$

   b. $7 \cdot d \cdot e$

   c. $5 \cdot 2 \cdot 2 \cdot y \cdot z$

   d. $3 \cdot 3 \cdot 2 \cdot 5 \cdot d$

2. Write the following expressions in expanded form.

   a. $3g$

   b. $11mp$

   c. $20yz$

   d. $15abc$

3. Find the product.

   a. $5d \cdot 7g$

   b. $12ab \cdot 3cd$

Number Correct: _____

Greatest Common Factor—Round 1

Directions: Determine the greatest common factor of each pair of numbers.

1.	GCF of 10 and 50	
2.	GCF of 5 and 35	
3.	GCF of 3 and 12	
4.	GCF of 8 and 20	
5.	GCF of 15 and 35	
6.	GCF of 10 and 75	
7.	GCF of 9 and 30	
8.	GCF of 15 and 33	
9.	GCF of 12 and 28	
10.	GCF of 16 and 40	
11.	GCF of 24 and 32	
12.	GCF of 35 and 49	
13.	GCF of 45 and 60	
14.	GCF of 48 and 72	
15.	GCF of 50 and 42	

16.	GCF of 45 and 72	
17.	GCF of 28 and 48	
18.	GCF of 44 and 77	
19.	GCF of 39 and 66	
20.	GCF of 64 and 88	
21.	GCF of 42 and 56	
22.	GCF of 28 and 42	
23.	GCF of 13 and 91	
24.	GCF of 16 and 84	
25.	GCF of 36 and 99	
26.	GCF of 39 and 65	
27.	GCF of 27 and 87	
28.	GCF of 28 and 70	
29.	GCF of 26 and 91	
30.	GCF of 34 and 51	

Greatest Common Factor—Round 2

Number Correct: _____
Improvement: _____

Directions: Determine the greatest common factor of each pair of numbers.

1.	GCF of 20 and 80	
2.	GCF of 10 and 70	
3.	GCF of 9 and 36	
4.	GCF of 12 and 24	
5.	GCF of 15 and 45	
6.	GCF of 10 and 95	
7.	GCF of 9 and 45	
8.	GCF of 18 and 33	
9.	GCF of 12 and 32	
10.	GCF of 16 and 56	
11.	GCF of 40 and 72	
12.	GCF of 35 and 63	
13.	GCF of 30 and 75	
14.	GCF of 42 and 72	
15.	GCF of 30 and 28	

16.	GCF of 33 and 99	
17.	GCF of 38 and 76	
18.	GCF of 26 and 65	
19.	GCF of 39 and 48	
20.	GCF of 72 and 88	
21.	GCF of 21 and 56	
22.	GCF of 28 and 52	
23.	GCF of 51 and 68	
24.	GCF of 48 and 84	
25.	GCF of 21 and 63	
26.	GCF of 64 and 80	
27.	GCF of 36 and 90	
28.	GCF of 28 and 98	
29.	GCF of 39 and 91	
30.	GCF of 38 and 95	

Example 1

a. Use the model to answer the following questions.

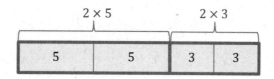

How many fives are in the model?

How many threes are in the model?

What does the expression represent in words?

What expression could we write to represent the model?

b. Use the new model and the previous model to answer the next set of questions.

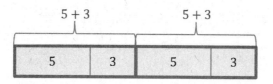

How many fives are in the model?

How many threes are in the model?

What does the expression represent in words?

What expression could we write to represent the model?

c.   Is the model in part (a) equivalent to the model in part (b)?

d.   What relationship do we see happening on either side of the equal sign?

e.   In Grade 5 and in Module 2 of this year, you have used similar reasoning to solve problems. What is the name of the property that is used to say that $2(5 + 3)$ is the same as $2 \times 5 + 2 \times 3$?

Example 2

Now we will take a look at an example with variables. Discuss the questions with your partner.

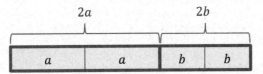

What does the model represent in words?

What does $2a$ mean?

How many a's are in the model?

How many b's are in the model?

What expression could we write to represent the model?

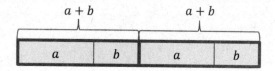

How many a's are in the expression?

How many b's are in the expression?

What expression could we write to represent the model?

Are the two expressions equivalent?

Example 3

Use GCF and the distributive property to write equivalent expressions.

1. $3f + 3g =$ _____

    What is the question asking us to do?

    How would Problem 1 look if we expanded each term?

    What is the GCF in Problem 1?

    How can we use the GCF to rewrite this expression?

2. $6x + 9y =$ _____

What is the question asking us to do?

How would Problem 2 look if we expanded each term?

What is the GCF in Problem 2?

How can we use the GCF to rewrite this expression?

3. $3c + 11c =$ _____

Is there a greatest common factor in Problem 3?

Rewrite the expression using the distributive property.

4. $24b + 8 =$ _____

Explain how you used GCF and the distributive property to rewrite the expression in Problem 4.

Why is there a 1 in the parentheses?

How is this related to the first two examples?

Lesson 11:   Factoring Expressions

Exercises

1. Apply the distributive property to write equivalent expressions.

   a. $7x + 7y$

   b. $15g + 20h$

   c. $18m + 42n$

   d. $30a + 39b$

   e. $11f + 15f$

   f. $18h + 13h$

   g. $55m + 11$

   h. $7 + 56y$

2. Evaluate each of the expressions below.

   a. $6x + 21y$ and $3(2x + 7y)$ $x = 3$ and $y = 4$

   b. $5g + 7g$ and $g(5 + 7)$ $g = 6$

c. $14x + 2$ and $2(7x + 1)$ $x = 10$

d.   Explain any patterns that you notice in the results to parts (a)–(c).

e.   What would happen if other values were given for the variables?

Closing

How can you use your knowledge of GCF and the distributive property to write equivalent expressions?

Find the missing value that makes the two expressions equivalent.

$4x + 12y$ _____$(x + 3y)$

$35x + 50y$ _____$(7x + 10y)$

$18x + 9y$ _____$(2x + y)$

$32x + 8y$ _____$(4x + y)$

$100x + 700y$ _____$(x + 7y)$

Explain how you determine the missing number.

Lesson 11:   Factoring Expressions

> **Lesson Summary**
>
> **AN EXPRESSION IN FACTORED FORM**: An expression that is a product of two or more expressions is said to be in *factored form*.

Name _____   Date _____

Use greatest common factor and the distributive property to write equivalent expressions in factored form.

1. $2x + 8y$

2. $13ab + 15ab$

3. $20g + 24h$

Name _____   Date _____

Use greatest common factor and the distributive property to write equivalent expressions in factored form.

1. $2x + 8y$

2. $13ab + 15ab$

3. $20g + 24h$

1. Use models to prove that $4(x + y)$ is equivalent to $4x + 4y$.

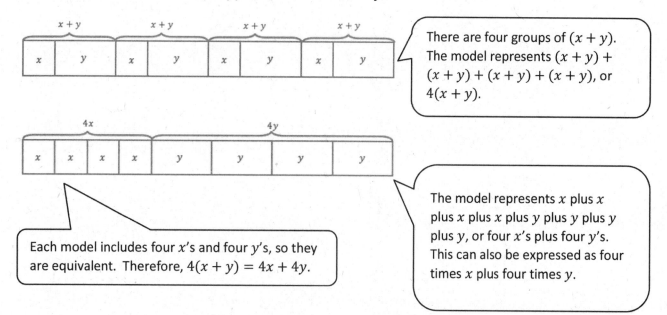

There are four groups of $(x + y)$. The model represents $(x + y) + (x + y) + (x + y) + (x + y)$, or $4(x + y)$.

The model represents x plus x plus x plus x plus y plus y plus y plus y, or four x's plus four y's. This can also be expressed as four times x plus four times y.

Each model includes four x's and four y's, so they are equivalent. Therefore, $4(x + y) = 4x + 4y$.

2. Use greatest common factor and the distributive property to write equivalent expressions in factored form for the following expressions.

   a. $4d + 12e$

   $\mathbf{4(d + 3e)}$ **or** $\mathbf{4(1d + 3e)}$

   b. $18x + 30y$

   $\mathbf{6(3x + 5y)}$

I can rewrite the expression as an equivalent expression in factored form, which means the expression is written as the product of factors. The number outside of the parentheses is the greatest common factor, or GCF.

The greatest common factor of $4d + 12e$ is 4, so I can write 4 outside of the parentheses.

1. Use models to prove that $3(a + b)$ is equivalent to $3a + 3b$.

2. Use greatest common factor and the distributive property to write equivalent expressions in factored form for the following expressions.

    a. $4d + 12e$

    b. $18x + 30y$

    c. $21a + 28y$

    d. $24f + 56g$

Opening Exercise

a. Create a model to show 2×5.

b. Create a model to show $2 \times b$, or $2b$.

Example 1

Write an expression that is equivalent to $2(a + b)$.

Create a model to represent $(a + b)$.

The expression $2(a + b)$ tells us that we have 2 of the $(a + b)$'s. Create a model that shows 2 groups of $(a + b)$.

How many a's and how many b's do you see in the diagram?

How would the model look if we grouped together the a's and then grouped together the b's?

What expression could we write to represent the new diagram?

What conclusion can we draw from the models about equivalent expressions?

Let $a = 3$ and $b = 4$.

What happens when we double $(a + b)$?

Example 2

Write an expression that is equivalent to double $(3x + 4y)$.

How can we rewrite double $(3x + 4y)$?

Is this expression in factored form, expanded form, or neither?

Let's start this problem the same way that we started the first example. What should we do?

How can we change the model to show $2(3x + 4y)$?

Are there terms that we can combine in this example?

What is an equivalent expression that we can use to represent $2(3x + 4y)$?

Summarize how you would solve this question without the model.

Example 3

Write an expression in expanded form that is equivalent to the model below.

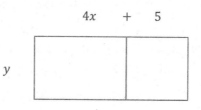

What factored expression is represented in the model?

How can we rewrite this expression in expanded form?

Example 4

Write an expression in expanded form that is equivalent to $3(7d + 4e)$.

Exercises

Create a model for each expression below. Then, write another equivalent expression using the distributive property.

1. $3(x + y)$

2. $4(2h + g)$

Apply the distributive property to write equivalent expressions in expanded form.

3. $8(h + 3)$

4. $3(2h + 7)$

5. $5(3x + 9y)$

6. $4(11h + 3g)$

7.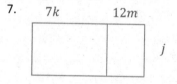

8. $a(9b + 13)$

Name _____ Date _____

Use the distributive property to write the following expressions in expanded form.

1. $2(b + c)$

2. $5(7h + 3m)$

3. $e(f + g)$

Name _____   Date _____

Use the distributive property to write the following expressions in expanded form.

1. $2(b + c)$

2. $5(7h + 3m)$

3. $e(f + g)$

1. Use the distributive property to write the following expressions in standard form.

    a. $9(x + y)$

        $9x + 9y$

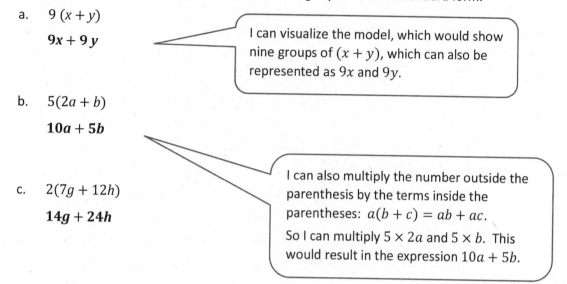

I can visualize the model, which would show nine groups of $(x + y)$, which can also be represented as $9x$ and $9y$.

    b. $5(2a + b)$

        $10a + 5b$

    c. $2(7g + 12h)$

        $14g + 24h$

I can also multiply the number outside the parenthesis by the terms inside the parentheses: $a(b + c) = ab + ac$.

So I can multiply $5 \times 2a$ and $5 \times b$. This would result in the expression $10a + 5b$.

2. Create a model to show that $3(2x + 3y) = 6x + 9y$.

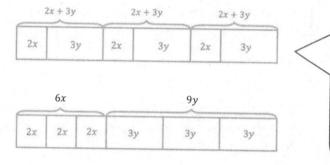

In the first model, there are three groups of $(2x + 3y)$. In the second model, there are three groups of $2x$, or $6x$ altogether, and three groups of $3y$, or a total of $9y$. In both models, there are three $2x$ terms and three $3y$ terms. They are just grouped differently.

1. Use the distributive property to write the following expressions in expanded form.

   a. $4(x + y)$

   b. $8(a + 3b)$

   c. $3(2x + 11y)$

   d. $9(7a + 6b)$

   e. $c(3a + b)$

   f. $y(2x + 11z)$

2. Create a model to show that $2(2x + 3y) = 4x + 6y$.

Example 1

Write an expression showing $1 \div 2$ without the use of the division symbol.

What can we determine from the model?

Example 2

Write an expression showing $a \div 2$ without the use of the division symbol.

What can we determine from the model?

When we write division expressions using the division symbol, we represent _____.

How would this look when we write division expressions using a fraction?

Example 3

   a.   Write an expression showing $a \div b$ without the use of the division symbol.

   b.   Write an expression for g divided by the quantity h plus 3.

   c.   Write an expression for the quotient of the quantity m reduced by 3 and 5.

Exercises

Write each expression two ways: using the division symbol and as a fraction.

   a.   12 divided by 4

   b.   3 divided by 5

   c. a divided by 4

   d.   The quotient of 6 and m

   e.   Seven divided by the quantity x plus y

   f. y divided by the quantity x minus 11

   g.   The sum of the quantity h and 3 divided by 4

   h.   The quotient of the quantity k minus 10 and m

Name _____   Date _____

Rewrite the expressions using the division symbol and as a fraction.

1. The quotient of m and 7

2. Five divided by the sum of a and b

3. The quotient of k decreased by 4 and 9

1. Rewrite the expressions using the division symbol and as a fraction.

    a. Eighteen divided by 23

    $18 \div 23$ *and* $\dfrac{18}{23}$

    > Writing a fraction to show division is more efficient than drawing models, arrays, or using the division symbol.

    b. The quotient of n and 9

    $n \div 9$ *and* $\dfrac{n}{9}$

    c. 8 divided by the sum of y and 5

    $8 \div (y+5)$ *and* $\dfrac{8}{y+5}$

    > When using the division symbol, I can show the sum of y and 5 by placing them in parentheses. I do not always need the parentheses in the denominator when writing the expression as a fraction.

2. Draw a model to show that $x \div 4$ is the same as $\dfrac{x}{4}$.

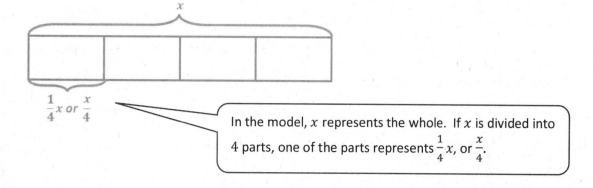

$\dfrac{1}{4}x$ *or* $\dfrac{x}{4}$

> In the model, x represents the whole. If x is divided into 4 parts, one of the parts represents $\dfrac{1}{4}x$, or $\dfrac{x}{4}$.

1. Rewrite the expressions using the division symbol and as a fraction.

   a. Three divided by 4

   b. The quotient of m and 11

   c. 4 divided by the sum of h and 7

   d. The quantity x minus 3 divided by y

2. Draw a model to show that $x \div 3$ is the same as $\dfrac{x}{3}$.

Example 1

Fill in the three remaining squares so that all the squares contain equivalent expressions.

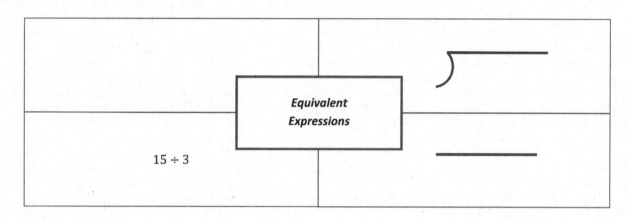

$15 \div 3$

Example 2

Fill in a blank copy of the four boxes using the words *dividend* and *divisor* so that it is set up for any example.

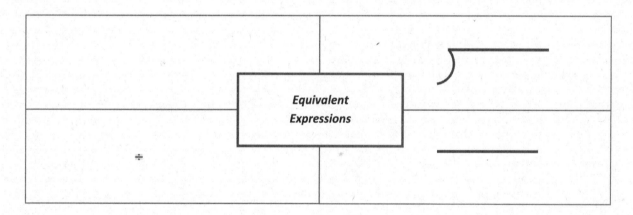

$\div$

Exercises

Complete the missing spaces in each rectangle set.

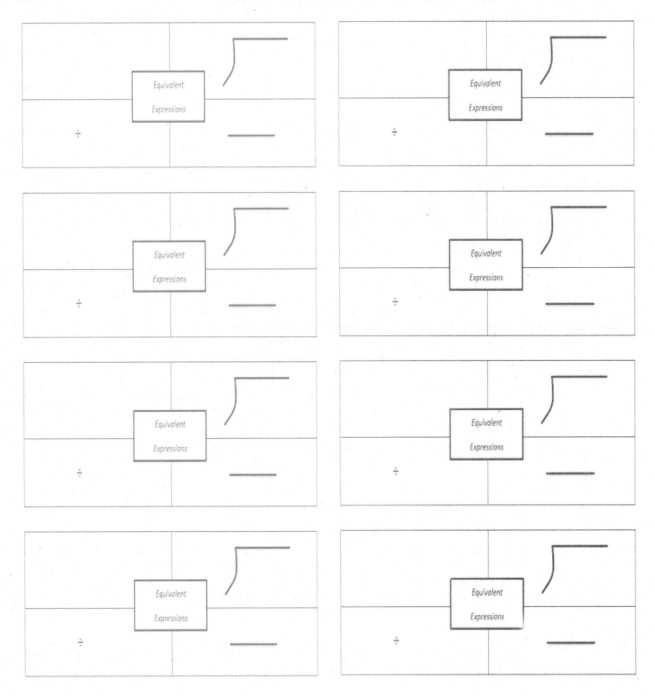

Lesson 14:   Writing Division Expressions

EUREKA
MATH®

Name _____   Date _____

1. Write the division expression in words and as a fraction.

$$(g + 12) \div h$$

2. Write the following division expression using the division symbol and as a fraction: f divided by the quantity h minus 3.

Complete the missing spaces in each rectangle set.

Since x is being divided by the sum of y and 12, it is the dividend. The divisor is $y + 12$ because it is what I am dividing by.

x divided by the sum of y and 12

$$y + 12 \overline{)\, x}$$

Equivalent Expressions

$$x \div (y + 12)$$

$$\frac{x}{y + 12}$$

Here, the dividend, x, is inside the division symbol because it is what is being divided. The divisor, $y + 12$, is outside the symbol because it is what I am dividing by.

Division can be represented as a fraction. The numerator represents the dividend, x. The denominator represents the divisor, $y + 12$.

Complete the missing spaces in each rectangle set.

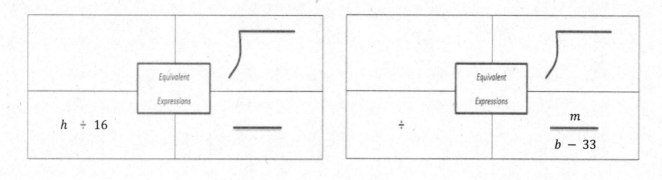

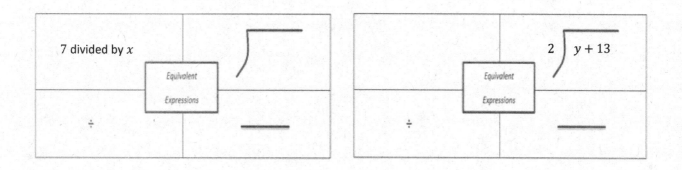

Opening Exercise

Complete the graphic organizer with mathematical words that indicate each operation. Some words may indicate more than one operation.

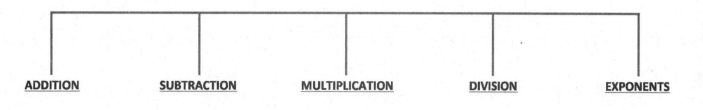

| ADDITION | SUBTRACTION | MULTIPLICATION | DIVISION | EXPONENTS |

Example 1

Write an expression using words.

    a. $a - b$

    b. xy

c. $4f + p$

d. $d - b^3$

e. $5(u - 10) + h$

f. $\dfrac{3}{d + f}$

Exercises

Circle all the vocabulary words that could be used to describe the given expression.

1. $6h - 10$

| ADDITION | SUBTRACTION | MULTIPLICATION | DIVISION |

2. $\dfrac{5d}{6}$

| SUM | DIFFERENCE | PRODUCT | QUOTIENT |

3. $5(2 + d) - 8$

| ADD | SUBTRACT | MULTIPLY | DIVIDE |

4. abc

| MORE THAN | LESS THAN | TIMES | EACH |

Write an expression using vocabulary to represent each given expression.

5. $8 - 2g$

6. $15(a + c)$

7. $\dfrac{m + n}{5}$

8. $b^3 - 18$

9. $f - \dfrac{d}{2}$

10. $\dfrac{u}{x}$

Name _____   Date _____

1. Write two word expressions for each problem using different math vocabulary for each expression.

    a. $5d - 10$

    b. $\dfrac{a}{b+2}$

2. List five different math vocabulary words that could be used to describe each given expression.

    a. $3(d - 2) + 10$

    b. $\dfrac{ab}{c}$

1. List five different vocabulary words that could be used to describe the given expression.

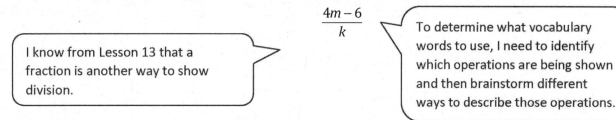

$$\frac{4m - 6}{k}$$

I know from Lesson 13 that a fraction is another way to show division.

To determine what vocabulary words to use, I need to identify which operations are being shown and then brainstorm different ways to describe those operations.

Possible Answers: Difference, less than, quotient, product, quadruple

2. Write an expression using math vocabulary for each expression below.

    a. $7 - 5h$

I need to show that 5 is being multiplied by h before it is subtracted from 7.

Possible Answers: The product of 5 and h subtracted from 7; seven minus the quantity 5 times h

    b. $\dfrac{m + 2}{4}$

I need to show that the entire numerator is being divided by 4. I can group m and 2 together using words like "the quantity."

Possible Answers: The quantity m plus 2 divided by 4; the quotient of m plus 2 and 4

1. List five different vocabulary words that could be used to describe each given expression.

   a. $a - d + c$

   b. $20 - 3c$

   c. $\dfrac{b}{d + 2}$

2. Write an expression using math vocabulary for each expression below.

   a. $5b - 18$

   b. $\dfrac{n}{2}$

   c. $a + (d - 6)$

   d. $10 + 2b$

Opening Exercise

Underline the key words in each statement.

    a.   The sum of twice b and 5

    b.   The quotient of c and d

    c. a raised to the fifth power and then increased by the product of 5 and c

    d.   The quantity of a plus b divided by 4

    e.   10 less than the product of 15 and c

    f.   5 times d and then increased by 8

Mathematical Modeling Exercise 1

Model how to change the expressions given in the Opening Exercise from words to variables and numbers.

    a.   The sum of twice b and 5

    b.   The quotient of c and d

    c. a raised to the fifth power and then increased by the product of 5 and c

    d.   The quantity of a plus b divided by 4

Lesson 16:   Write Expressions in Which Letters Stand for Numbers   **179**

e.   10 less than the product of 15 and c

f.   5 times d and then increased by 8

Mathematical Modeling Exercise 2

Model how to change each real-world scenario to an expression using variables and numbers. Underline the text to show the key words before writing the expression.

Marcus has 4 more dollars than Yaseen. If y is the amount of money Yaseen has, write an expression to show how much money Marcus has.

Mario is missing half of his assignments. If a represents the number of assignments, write an expression to show how many assignments Mario is missing.

Kamilah's weight has tripled since her first birthday. If w represents the amount Kamilah weighed on her first birthday, write an expression to show how much Kamilah weighs now.

Nathan brings cupcakes to school and gives them to his five best friends, who share them equally. If c represents the number of cupcakes Nathan brings to school, write an expression to show how many cupcakes each of his friends receive.

Mrs. Marcus combines her atlases and dictionaries and then divides them among 10 different tables. If a represents the number of atlases and d represents the number of dictionaries Mrs. Marcus has, write an expression to show how many books would be on each table.

To improve in basketball, Ivan's coach told him that he needs to take four times as many free throws and four times as many jump shots every day. If f represents the number of free throws and j represents the number of jump shots Ivan shoots daily, write an expression to show how many shots he will need to take in order to improve in basketball.

Exercises

Mark the text by underlining key words, and then write an expression using variables and/or numbers for each statement.

1. b decreased by c squared

2. 24 divided by the product of 2 and a

3. 150 decreased by the quantity of 6 plus b

4. The sum of twice c and 10

5. Marlo had $35 but then spent $\$m$.

6. Samantha saved her money and was able to quadruple the original amount, m.

7. Veronica increased her grade, g, by 4 points and then doubled it.

8. Adbell had m pieces of candy and ate 5 of them. Then, he split the remaining candy equally among 4 friends.

9. To find out how much paint is needed, Mr. Jones must square the side length, s, of the gate and then subtract 15.

10. Luis brought x cans of cola to the party, Faith brought d cans of cola, and De'Shawn brought h cans of cola. How many cans of cola did they bring altogether?

Name _____   Date _____

Mark the text by underlining key words, and then write an expression using variables and/or numbers for each of the statements below.

1. Omaya picked x amount of apples, took a break, and then picked v more. Write the expression that models the total number of apples Omaya picked.

2. A number h is tripled and then decreased by 8.

3. Sidney brought s carrots to school and combined them with Jenan's j carrots. She then splits them equally among 8 friends.

4. 15 less than the quotient of e and d

5. Marissa's hair was 10 inches long, and then she cut h inches.

Mark the text by underlining key words, and then write an expression using variables and numbers for each of the statements below.

1. The difference of g and 18 is divided by h squared.

    The <u>difference</u> of g and 18 is <u>divided</u> by h <u>squared</u>.

    > I need to determine the operations that the key words are describing. *Difference* is the result of subtraction.

    $$\frac{g-18}{h^2}$$

    > I can write "h squared" as h to the second power.

2. Noelle read p pages yesterday. Marcus read 9 pages more than one-third of the pages Noelle read. Write an expression that represents the number of pages that Marcus read.

    Noelle read p pages yesterday. Marcus read 9 pages <u>more than</u> <u>one-third</u> of the pages Noelle read. Write an expression that represents the number of pages that Marcus read.

    $\frac{1}{3}p + 9$ *or* $\frac{p}{3} + 9$ *or* $p \div 3 + 9$

    > To determine the number of pages Marcus read, I need to represent one-third of the pages Noelle read before I can add 9 to that amount.

    > *One-third of* describes multiplication. Because $\frac{1}{3}$ is a fraction, it also describes division.

Mark the text by underlining key words, and then write an expression using variables and numbers for each of the statements below.

1. Justin can type w words per minute. Melvin can type 4 times as many words as Justin. Write an expression that represents the rate at which Melvin can type.

2. Yohanna swam y yards yesterday. Sheylin swam 5 yards less than half the amount of yards as Yohanna. Write an expression that represents the number of yards Sheylin swam yesterday.

3. A number d is decreased by 5 and then doubled.

4. Nahom had n baseball cards, and Semir had s baseball cards. They combined their baseball cards and then sold 10 of them.

5. The sum of 25 and h is divided by f cubed.

Number Correct: _____

Addition of Decimals I—Round 1

Directions: Evaluate each expression.

1.	5.1 + 6	
2.	5.1 + 0.6	
3.	5.1 + 0.06	
4.	5.1 + 0.006	
5.	5.1 + 0.0006	
6.	3 + 2.4	
7.	0.3 + 2.4	
8.	0.03 + 2.4	
9.	0.003 + 2.4	
10.	0.0003 + 2.4	
11.	24 + 0.3	
12.	2 + 0.3	
13.	0.2 + 0.03	
14.	0.02 + 0.3	
15.	0.2 + 3	
16.	2 + 0.03	
17.	5 + 0.4	
18.	0.5 + 0.04	
19.	0.05 + 0.4	
20.	0.5 + 4	
21.	5 + 0.04	
22.	0.5 + 0.4	

23.	3.6 + 2.1	
24.	3.6 + 0.21	
25.	3.6 + 0.021	
26.	0.36 + 0.021	
27.	0.036 + 0.021	
28.	1.4 + 42	
29.	1.4 + 4.2	
30.	1.4 + 0.42	
31.	1.4 + 0.042	
32.	0.14 + 0.042	
33.	0.014 + 0.042	
34.	0.8 + 2	
35.	0.8 + 0.2	
36.	0.08 + 0.02	
37.	0.008 + 0.002	
38.	6 + 0.4	
39.	0.6 + 0.4	
40.	0.06 + 0.04	
41.	0.006 + 0.004	
42.	0.1 + 9	
43.	0.1 + 0.9	
44.	0.01 + 0.09	

Addition of Decimals I—Round 2

Number Correct: _____
Improvement: _____

Directions: Evaluate each expression.

1.	3.2 + 5		23.	4.2 + 5.5	
2.	3.2 + 0.5		24.	4.2 + 0.55	
3.	3.2 + 0.05		25.	4.2 + 0.055	
4.	3.2 + 0.005		26.	0.42 + 0.055	
5.	3.2 + 0.0005		27.	0.042 + 0.055	
6.	4 + 5.3		28.	2.7 + 12	
7.	0.4 + 5.3		29.	2.7 + 1.2	
8.	0.04 + 5.3		30.	2.7 + 0.12	
9.	0.004 + 5.3		31.	2.7 + 0.012	
10.	0.0004 + 5.3		32.	0.27 + 0.012	
11.	4 + 0.53		33.	0.027 + 0.012	
12.	6 + 0.2		34.	0.7 + 3	
13.	0.6 + 0.02		35.	0.7 + 0.3	
14.	0.06 + 0.2		36.	0.07 + 0.03	
15.	0.6 + 2		37.	0.007 + 0.003	
16.	2 + 0.06		38.	5 + 0.5	
17.	1 + 0.7		39.	0.5 + 0.5	
18.	0.1 + 0.07		40.	0.05 + 0.05	
19.	0.01 + 0.7		41.	0.005 + 0.005	
20.	0.1 + 7		42.	0.2 + 8	
21.	1 + 0.07		43.	0.2 + 0.8	
22.	0.1 + 0.7		44.	0.02 + 0.08	

Exercises

Station One	1. The sum of *a* and *b*
	2. Five more than twice a number *c*
	3. Martha bought *d* number of apples and then ate 6 of them.
Station Two	1. 14 decreased by *p*
	2. The total of *d* and *f*, divided by 8
	3. Rashod scored 6 less than 3 times as many baskets as Mike. Mike scored *b* baskets.
Station Three	1. The quotient of *c* and 6
	2. Triple the sum of *x* and 17
	3. Gabrielle had *b* buttons but then lost 6. Gabrielle took the remaining buttons and split them equally among her 5 friends.

Station Four	1.	*d* doubled
	2.	Three more than 4 times a number *x*
	3.	Mali has *c* pieces of candy. She doubles the amount of candy she has and then gives away 15 pieces.
Station Five	1.	*f* cubed
	2.	The quantity of 4 increased by *a*, and then the sum is divided by 9.
	3.	Tai earned 4 points fewer than double Oden's points. Oden earned *p* points.
Station Six	1.	The difference between *d* and 8
	2.	6 less than the sum of *d* and 9
	3.	Adalyn has *x* pants and *s* shirts. She combined them and sold half of them. How many items did Adalyn sell?

Lesson 17:   Write Expressions in Which Letters Stand for Numbers

Name _____   Date _____

Write an expression using letters and/or numbers for each problem below.

1. *d* squared

2. A number *x* increased by 6, and then the sum is doubled.

3. The total of *h* and *b* is split into 5 equal groups.

4. Jazmin has increased her $45 by *m* dollars and then spends a third of the entire amount.

5. Bill has *d* more than 3 times the number of baseball cards as Frank. Frank has *f* baseball cards.

Write an expression using letters and/or numbers for each problem below.

1. 13 times the difference of k and 7

    $13(k - 7)$

    > I can use parentheses to show that 13 is being multiplied by the difference of k and 7 instead of just being multiplied by k.

2. The quantity of h increased by 14 divided by three times m

    $\dfrac{h + 14}{3m}$

    > I can show the quotient using a fraction. The first quantity will be the numerator because that is the dividend (what is being divided), and the second quantity, which I am dividing by (or the divisor), will be the denominator.

3. Melinda can do 2 times as many push-ups as Tim and Quinn combined. Tim can do t push-ups, and Quinn can do q push-ups.

    $2(t + q)$

    > *Tim and Quinn combined* means that I need to add together the number of push-ups that each of them can do before multiplying by two.

4. Yesterday, the temperature was 28 degrees warmer than triple the temperature, t, four months ago.

    $3t + 28$

    > *Warmer* tells me I need to add 28 to the product of t and 3.

Write an expression using letters and/or numbers for each problem below.

1. 4 less than the quantity of 8 times *n*

2. 6 times the sum of *y* and 11

3. The square of *m* reduced by 49

4. The quotient when the quantity of 17 plus *p* is divided by 8

5. Jim earned *j* in tips, and Steve earned *s* in tips. They combine their tips and then split them equally.

6. Owen has *c* collector cards. He quadruples the number of cards he has and then combines them with Ian, who has *i* collector cards.

7. Rae runs 4 times as many miles as Madison and Aaliyah combined. Madison runs *m* miles, and Aaliyah runs *a* miles.

8. By using coupons, Mary Jo is able to decrease the retail price of her groceries, *g*, by $125.

9. To calculate the area of a triangle, you find the product of the base and height and then divide by 2.

10. The temperature today was 10 degrees colder than twice yesterday's temperature, *t*.

Opening Exercise

How can we show a number increased by 2?

Can you prove this using a model?

Example 1: The Importance of Being Specific in Naming Variables

When naming variables in expressions, it is important to be very clear about what they represent. The units of measure must be included if something is measured.

Exercises 1–2

1. Read the variable in the table, and improve the description given, making it more specific.

Variable	Incomplete Description	Complete Description with Units
Joshua's speed (J)	Let J represent Joshua's speed.	
Rufus's height (R)	Let R represent Rufus's height.	
Milk sold (M)	Let M represent the amount of milk sold.	
Colleen's time in the 40-meter hurdles (C)	Let C represent Colleen's time.	
Sean's age (S)	Let S represent Sean's age.	

2.   Read each variable in the table, and improve the description given, making it more specific.

Variable	Incomplete Description	Complete Description with Units
Karolyn's CDs (K)	Let K represent Karolyn's CDs.	Let K represent the number of CDs Karolyn has.
Joshua's merit badges (J)	Let J represent Joshua's merit badges.	
Rufus's trading cards (R)	Let R represent Rufus's trading cards.	
Milk money (M)	Let M represent the amount of milk money.	

Example 2: Writing and Evaluating Addition and Subtraction Expressions

Read each story problem. Identify the unknown quantity, and write the addition or subtraction expression that is described. Finally, evaluate your expression using the information given in column four.

Story Problem	Description with Units	Expression	Evaluate the Expression If:	Show Your Work and Evaluate
Gregg has two more dollars than his brother Jeff. Write an expression for the amount of money Gregg has.	Let j represent Jeff's money in dollars.	$j + 2$	Jeff has $12.	$j + 2$ $12 + 2$ 14 Gregg has $14.
Gregg has two more dollars than his brother Jeff. Write an expression for the amount of money Jeff has.	Let g represent Gregg's money in dollars.	$g - 2$	Gregg has $14.	$g - 2$ $14 - 2$ 12 Jeff has $12.
Abby read 8 more books than Kristen in the first marking period. Write an expression for the number of books Abby read.			Kristen read 9 books in the first marking period.	

Lesson 18:   Writing and Evaluating Expressions—Addition and Subtraction

Abby read 6 more books than Kristen in the second marking period. Write an expression for the number of books Kristen read.			Abby read 20 books in the second marking period.	
Daryl has been teaching for one year longer than Julie. Write an expression for the number of years that Daryl has been teaching.			Julie has been teaching for 28 years.	
Ian scored 4 fewer goals than Julia in the first half of the season. Write an expression for the number of goals Ian scored.			Julia scored 13 goals.	
Ian scored 3 fewer goals than Julia in the second half of the season. Write an expression for the number of goals Julia scored.			Ian scored 8 goals.	
Johann visited Niagara Falls 3 times fewer than Arthur. Write an expression for the number of times Johann visited Niagara Falls.			Arthur visited Niagara Falls 5 times.	

Name _____   Date _____

Kathleen lost a tooth today. Now she has lost 4 more than her sister Cara lost.

1. Write an expression to represent the number of teeth Cara has lost. Let K represent the number of teeth Kathleen lost.

    Expression:

2. Write an expression to represent the number of teeth Kathleen has lost. Let C represent the number of teeth Cara lost.

    Expression:

3. If Cara lost 3 teeth, how many teeth has Kathleen lost?

1. Read the story problem. Part (a): Identify the unknown quantity, and write the addition or subtraction expression that is described. Part (b): Evaluate your expression using the information given.

   a. The home football team scored 17 more points than the away team.

   Description with units: Let p represent the points the away team scored.

   Expression: p + 17

   > I do not know how many points the away team scored, so I will use a variable. Then I can add 17 to determine the points the home team scored.

   b. The away team scored 19 points in the game.

   $$p + 17$$
   $$19 + 17$$
   $$36$$

   The home team scored 36 points in the game.

2. If the home team had scored 42 points, how would you determine the number of points scored by the away team?

   I would subtract 17 points from 42 points to get 25 points for the away team.

   > This time I was given the points the home team scored instead of the points the away team scored. So I need to do the opposite of adding.

1. Read each story problem. Identify the unknown quantity, and write the addition or subtraction expression that is described. Finally, evaluate your expression using the information given in column four.

Story Problem	Description with Units	Expression	Evaluate the Expression If:	Show Your Work and Evaluate
Sammy has two more baseballs than his brother Ethan.	Let *e* represent the number of balls Ethan has.	$e + 2$	Ethan has 7 baseballs.	$e + 2$ $7 + 2$ 9 Sammy has 9 baseballs.
Ella wrote 8 more stories than Anna in the fifth grade.			Anna wrote 10 stories in the fifth grade.	
Lisa has been dancing for 3 more years than Danika.			Danika has been dancing for 6 years.	
The New York Rangers scored 2 fewer goals than the Buffalo Sabres last night.			The Rangers scored 3 goals last night.	
George has gone camping 3 times fewer than Dave.			George has gone camping 8 times.	

2. If George went camping 15 times, how could you figure out how many times Dave went camping?

Subtraction of Decimals—Round 1

Directions: Evaluate each expression.

1.	55 – 50	
2.	55 – 5	
3.	5.5 – 5	
4.	5.5 – 0.5	
5.	88 – 80	
6.	88 – 8	
7.	8.8 – 8	
8.	8.8 – 0.8	
9.	33 – 30	
10.	33 – 3	
11.	3.3 – 3	
12.	1 – 0.3	
13.	1 – 0.03	
14.	1 – 0.003	
15.	0.1 – 0.03	
16.	4 – 0.8	
17.	4 – 0.08	
18.	4 – 0.008	
19.	0.4 – 0.08	
20.	9 – 0.4	
21.	9 – 0.04	
22.	9 – 0.004	

23.	9.9 – 5	
24.	9.9 – 0.5	
25.	0.99 – 0.5	
26.	0.99 – 0.05	
27.	4.7 – 2	
28.	4.7 – 0.2	
29.	0.47 – 0.2	
30.	0.47 – 0.02	
31.	8.4 – 1	
32.	8.4 – 0.1	
33.	0.84 – 0.1	
34.	7.2 – 5	
35.	7.2 – 0.5	
36.	0.72 – 0.5	
37.	0.72 – 0.05	
38.	8.6 – 7	
39.	8.6 – 0.7	
40.	0.86 – 0.7	
41.	0.86 – 0.07	
42.	5.1 – 4	
43.	5.1 – 0.4	
44.	0.51 – 0.4	

Number Correct: _____

Subtraction of Decimals—Round 2

Number Correct: _____
Improvement: _____

Directions: Evaluate each expression.

1.	66 – 60	
2.	66 – 6	
3.	6.6 – 6	
4.	6.6 – 0.6	
5.	99 – 90	
6.	99 – 9	
7.	9.9 – 9	
8.	9.9 – 0.9	
9.	22 – 20	
10.	22 – 2	
11.	2.2 – 2	
12.	3 – 0.4	
13.	3 – 0.04	
14.	3 – 0.004	
15.	0.3 – 0.04	
16.	8 – 0.2	
17.	8 – 0.02	
18.	8 – 0.002	
19.	0.8 – 0.02	
20.	5 – 0.1	
21.	5 – 0.01	
22.	5 – 0.001	

23.	6.8 – 4	
24.	6.8 – 0.4	
25.	0.68 – 0.4	
26.	0.68 – 0.04	
27.	7.3 – 1	
28.	7.3 – 0.1	
29.	0.73 – 0.1	
30.	0.73 – 0.01	
31.	9.5 – 2	
32.	9.5 – 0.2	
33.	0.95 – 0.2	
34.	8.3 – 5	
35.	8.3 – 0.5	
36.	0.83 – 0.5	
37.	0.83 – 0.05	
38.	7.2 – 4	
39.	7.2 – 0.4	
40.	0.72 – 0.4	
41.	0.72 – 0.04	
42.	9.3 – 7	
43.	9.3 – 0.7	
44.	0.93 – 0.7	

EUREKA MATH®

Opening Exercise

My older sister is exactly two years older than I am. Sharing a birthday is both fun and annoying. Every year on our birthday, we have a party, which is fun, but she always brags that she is two years older than I am, which is annoying. Shown below is a table of our ages, starting when I was born:

My Age (in years)	My Sister's Age (in years)
0	2
1	3
2	4
3	5
4	6

a. Looking at the table, what patterns do you see? Tell a partner.

b. On the day I turned 8 years old, how old was my sister?

c. How do you know?

d. On the day I turned 16 years old, how old was my sister?

e. How do you know?

f. Do we need to extend the table to calculate these answers?

Example 1

My Age (in years)	My Sister's Age (in years)
0	2
1	3
2	4
3	5
4	6

a. What if you don't know how old I am? Let's use a variable for my age. Let $Y =$ my age in years. Can you develop an expression to describe how old my sister is?

b. Please add that to the last row of the table.

Example 2

My Age (in years)	My Sister's Age (in years)
0	2
1	3
2	4
3	5
4	6

a. How old was I when my sister was 6 years old?

b. How old was I when my sister was 15 years old?

c. How do you know?

Lesson 19:   Substituting to Evaluate Addition and Subtraction
Expressions

© 2019 Great Minds®. eureka-math.org

d.   Look at the table in Example 2. If you know my sister's age, can you determine my age?

e.   If we use the variable G for my sister's age in years, what expression would describe my age in years?

f.   Fill in the last row of the table with the expressions.

g.   With a partner, calculate how old I was when my sister was 22, 23, and 24 years old.

Exercises

1. Noah and Carter are collecting box tops for their school. They each bring in 1 box top per day starting on the first day of school. However, Carter had a head start because his aunt sent him 15 box tops before school began. Noah's grandma saved 10 box tops, and Noah added those on his first day.

a.   Fill in the missing values that indicate the total number of box tops each boy brought to school.

School Day	Number of Box Tops Noah Has	Number of Box Tops Carter Has
1	11	16
2		
3		
4		
5		

b.   If we let D be the number of days since the new school year began, on day D of school, how many box tops will Noah have brought to school?

c.   On day D of school, how many box tops will Carter have brought to school?

d.   On day 10 of school, how many box tops will Noah have brought to school?

e.   On day 10 of school, how many box tops will Carter have brought to school?

2. Each week the Primary School recycles 200 pounds of paper. The Intermediate School also recycles the same amount but had another 300 pounds left over from summer school. The Intermediate School custodian added this extra 300 pounds to the first recycle week.

a. Number the weeks, and record the amount of paper recycled by both schools.

Week	Total Amount of Paper Recycled by the Primary School This School Year in Pounds	Total Amount of Paper Recycled by the Intermediate School This School Year in Pounds

b. If this trend continues, what will be the total amount collected for each school on Week 10?

3. Shelly and Kristen share a birthday, but Shelly is 5 years older.

a. Make a table showing their ages every year, beginning when Kristen was born.

b. If Kristen is 16 years old, how old is Shelly?

c. If Kristen is K years old, how old is Shelly?

d. If Shelly is S years old, how old is Kristen?

Lesson 19: Substituting to Evaluate Addition and Subtraction Expressions

Name _____ Date _____

Jenna and Allie work together at a piano factory. They both were hired on January 3, but Jenna was hired in 2005, and Allie was hired in 2009.

a. Fill in the table below to summarize the two workers' experience totals.

Year	Allie's Years of Experience	Jenna's Years of Experience
2010		
2011		
2012		
2013		
2014		

b. If both workers continue working at the piano factory, when Allie has A years of experience on the job, how many years of experience will Jenna have on the job?

c. If both workers continue working at the piano factory, when Allie has 20 years of experience on the job, how many years of experience will Jenna have on the job?

1. Makenzie and Micah went to Caspersen Beach, Florida, to collect shark teeth. Before they went, Makenzie had 13 teeth in her collection, and Micah had 4 teeth in his collection. On an 8-day trip, they each collected 3 new teeth each day.

   a. Make a table showing how many teeth each person had in his or her collection at the end of each day.

Day	Number of Shark Teeth in Makenzie's Collection	Number of Shark Teeth in Micah's Collection
1	16	7
2	19	10
3	22	13
4	25	16
5	28	19
6	31	22
7	34	25
8	37	28

> On the first day, I need to add 3 to the totals for each person. And each day after that I will add 3 more to represent the new shark teeth they found.

> My table needs a row for each of the 8 days in the trip. It also needs a column for each person.

   b. If this pattern of shark teeth finding continues, how many teeth does Micah have when Makenzie has T shark teeth?

> I can see that Micah has fewer shark teeth. So I know I will be subtracting some amount from the number that Makenzie has.

   When Makenzie has T shark teeth, Micah has $T - 9$ shark teeth.

c. If this pattern of shark teeth finding continues, how many shark teeth does Micah have when Makenzie has 70 shark teeth?

> I need to use the expression I came up with in part (b) to help me answer the question.

$70 - 9 = 61$

When Makenzie has 70 shark teeth, Micah has 61 shark teeth.

d. If this pattern of shark teeth finding continues, how many teeth does Makenzie have when Micah has M shark teeth?

When Micah has M shark teeth, Makenzie has $M + 9$ shark teeth.

> I already know that the number of teeth is always 9 apart, but this time I need to add 9 because Makenzie has 9 more than Micah every day.

2. Maya and Albert are making necklaces that consist of large round beads and small square beads. The relationship between the number of large round beads and the total number of beads is shown in the table.

Number of Large Round Beads	Total Number of Beads
0	6
1	7
2	8
5	11
50	56

> In this problem, I am given a completed table. I need to see what number was added to the number of large round beads to determine the total number of beads.

a. Maya wrote an expression for the relationship depicted in the table as $R + 6$. Albert wrote an expression for the same relationship as $T - 6$. Is it possible to have two different expressions to represent one relationship? Explain.

Both expressions can represent the same relationship, depending on the point of view. The expression $R + 6$ represents the number of large round beads plus the number of small square beads. The expression $T - 6$ represents the number of small square beads taken away from the total number of beads.

     Lesson 19:   Substituting to Evaluate Addition and Subtraction Expressions

b. What do you think the variable in each student's expression represents? How would you define them?

The variable T would represent the total number of beads on the necklace. The variable R would represent the number of large round beads.

c. If the same pattern continues, how many large beads will be on the necklace if there are 72 beads total?

$$T - 6$$
$$72 - 6$$
$$66$$

Because I am given the total number of beads, it makes sense to use the expression $T - 6$ to solve for the number of large round beads.

There would be 66 large round beads used if there are 72 total beads.

1. Suellen and Tara are in sixth grade, and both take dance lessons at Twinkle Toes Dance Studio. This is Suellen's first year, while this is Tara's fifth year of dance lessons. Both girls plan to continue taking lessons throughout high school.

    a. Complete the table showing the number of years the girls will have danced at the studio.

Grade	Suellen's Years of Experience Dancing	Tara's Years of Experience Dancing
Sixth		
Seventh		
Eighth		
Ninth		
Tenth		
Eleventh		
Twelfth		

    b. If Suellen has been taking dance lessons for Y years, how many years has Tara been taking lessons?

2. Daejoy and Damian collect fossils. Before they went on a fossil-hunting trip, Daejoy had 25 fossils in her collection, and Damian had 16 fossils in his collection. On a 10-day fossil-hunting trip, they each collected 2 new fossils each day.

    a. Make a table showing how many fossils each person had in their collection at the end of each day.

    b. If this pattern of fossil finding continues, how many fossils does Damian have when Daejoy has F fossils?

    c. If this pattern of fossil finding continues, how many fossils does Damian have when Daejoy has 55 fossils?

3. A train consists of three types of cars: box cars, an engine, and a caboose. The relationship among the types of cars is demonstrated in the table below.

Number of Box Cars	Number of Cars in the Train
0	2
1	3
2	4
10	12
100	102

a. Tom wrote an expression for the relationship depicted in the table as $B + 2$. Theresa wrote an expression for the same relationship as $C - 2$. Is it possible to have two different expressions to represent one relationship? Explain.

b. What do you think the variable in each student's expression represents? How would you define them?

4. David was 3 when Marieka was born. Complete the table.

Marieka's Age in Years	David's Age in Years
5	8
6	9
7	10
8	11
10	
	20
32	
M	
	D

5. Caitlin and Michael are playing a card game. In the first round, Caitlin scored 200 points, and Michael scored 175 points. In each of the next few rounds, they each scored 50 points. Their score sheet is below.

Caitlin's Points	Michael's Points
200	175
250	225
300	275
350	325

a. If this trend continues, how many points will Michael have when Caitlin has 600 points?

b. If this trend continues, how many points will Michael have when Caitlin has C points?

c. If this trend continues, how many points will Caitlin have when Michael has 975 points?

d. If this trend continues, how many points will Caitlin have when Michael has M points?

Lesson 19:   Substituting to Evaluate Addition and Subtraction Expressions

6. The high school marching band has 15 drummers this year. The band director insists that there are to be 5 more trumpet players than drummers at all times.

   a. How many trumpet players are in the marching band this year?

   b. Write an expression that describes the relationship of the number of trumpet players (T) and the number of drummers (D).

   c. If there are only 14 trumpet players interested in joining the marching band next year, how many drummers will the band director want in the band?

Example 1

The farmers' market is selling bags of apples. In every bag, there are 3 apples.

a. Complete the table.

Number of Bags	Total Number of Apples
1	3
2	
3	
4	
B	

b. What if the market had 25 bags of apples to sell? How many apples is that in all?

c. If a truck arrived that had some number, *a*, more apples on it, then how many bags would the clerks use to bag up the apples?

d. If a truck arrived that had 600 apples on it, how many bags would the clerks use to bag up the apples?

e. How is part (d) different from part (b)?

Exercises 1–3

1. In New York State, there is a five-cent deposit on all carbonated beverage cans and bottles. When you return the empty can or bottle, you get the five cents back.

    a. Complete the table.

Number of Containers Returned	Refund in Dollars
1	
2	
3	
4	
10	
50	
100	
C	

    b. If we let C represent the number of cans, what is the expression that shows how much money is returned?

    c. Use the expression to find out how much money Brett would receive if he returned 222 cans.

    d. If Gavin needs to earn $4.50 for returning cans, how many cans does he need to collect and return?

    e. How is part (d) different from part (c)?

Lesson 20:   Writing and Evaluating Expressions—Multiplication and Division

© 2019 Great Minds®. eureka-math.org

EUREKA
MATH®

2. The fare for a subway or a local bus ride is $2.50.

   a. Complete the table.

Number of Rides	Cost of Rides in Dollars
1	
2	
3	
4	
5	
10	
30	
R	

   b. If we let R represent the number of rides, what is the expression that shows the cost of the rides?

   c. Use the expression to find out how much money 60 rides would cost.

   d. If a commuter spends $175.00 on subway or bus rides, how many trips did the commuter take?

   e. How is part (d) different from part (c)?

Challenge Problem

3. A pendulum swings though a certain number of cycles in a given time. Owen made a pendulum that swings 12 times every 15 seconds.

    a. Construct a table showing the number of cycles through which a pendulum swings. Include data for up to one minute. Use the last row for C cycles, and write an expression for the time it takes for the pendulum to make C cycles.

    b. Owen and his pendulum team set their pendulum in motion and counted 16 cycles. What was the elapsed time?

    c. Write an expression for the number of cycles a pendulum swings in S seconds.

    d. In a different experiment, Owen and his pendulum team counted the cycles of the pendulum for 35 seconds. How many cycles did they count?

Lesson 20:   Writing and Evaluating Expressions—Multiplication and Division

© 2019 Great Minds®. eureka-math.org

Name _____ Date _____

Anna charges $8.50 per hour to babysit. Complete the table, and answer the questions below.

Number of Hours	Amount Anna Charges in Dollars
1	
2	
5	
8	
H	

a. Write an expression describing her earnings for working H hours.

b. How much will she earn if she works for $3\frac{1}{2}$ hours?

c. How long will it take Anna to earn $51.00?

1. A seamstress can sew 180 skirts per month.

    a. Write an expression describing how many skirts are made by the seamstress in *M* months.

        180*M*

        > I can replace the variable with a number to help me think through the problem. If the seamstress sewed for 2 months, she would make 180×2 skirts. And in 3 months, she would make 180×3 skirts. So in *M* months she must make $180 \times M$ skirts.

    b. How many skirts will be in an entire year (12 months)?

        $180 \cdot 12 = 2,160$. *There will be 2,160 skirts made in a year.*

    c. How long does it take the seamstress to complete 1,980 skirts?

        $1,980 \text{ skirts} \div 180 \frac{\text{skirts}}{\text{month}} = 11 \text{ months}$

        > To get the total number of skirts, I multiplied. So to get the number of months, I will do the opposite, or inverse.

2. Malik is a hot dog vendor, for which he earns $2.30 per hot dog sold. Create a table of values that shows the relationship between the number of hot dogs that Malik sells, H, and the amount of money Malik earns in dollars, D.

> I can choose any numbers for the number of hot dogs, like $1, 2, 3, 4,$ or $10, 20, 30, 40,$ and then use them to determine the total earnings.

Number of Hot Dogs Sold (H)	Malik's Earnings in Dollars (D)
1	2.30
10	23.00
100	230.00
1,000	2,300.00

a. If you know how many hot dogs Malik sold, can you determine how much money he earned? Write the corresponding expression.

   Multiplying the number of hot dogs that Malik sold by his profit rate ($2.30 per hot dog) will calculate his money earned. 2.30H is the expression for his earnings in dollars.

b. Use your expression to determine how much Malik earned by selling 90 hot dogs.

   $2.30H = 2.30 \cdot 90 = 207.$ *Malik earned* $207.00 *for selling* 90 *hot dogs.*

   > I am given the number of hot dogs, which I can use to replace H in the expression from part (a), and multiply.

c. Malik must earn $1,334 each week to cover all of his expenses. Determine how many hot dogs Malik must sell in order to earn $1,334 in a week.

   $1,334 \div 2.30 = 580.$ *Therefore, Malik must sell* 580 *hot dogs each week in order to cover his expenses.*

   > I am given D, the amount of money in dollars. So I will need to use the opposite operation to solve for H, the number of hot dogs.

Lesson 20: Writing and Evaluating Expressions—Multiplication and Division

1. A radio station plays 12 songs each hour. They never stop for commercials, news, weather, or traffic reports.

   a. Write an expression describing how many songs are played by the radio station in H hours.

   b. How many songs will be played in an entire day (24 hours)?

   c. How long does it take the radio station to play 60 consecutive songs?

2. A ski area has a high-speed lift that can move 2,400 skiers to the top of the mountain each hour.

   a. Write an expression describing how many skiers can be lifted in H hours.

   b. How many skiers can be moved to the top of the mountain in 14 hours?

   c. How long will it take to move 3,600 skiers to the top of the mountain?

3. Polly writes a magazine column, for which she earns $35 per hour. Create a table of values that shows the relationship between the number of hours that Polly works, H, and the amount of money Polly earns in dollars, E.

   a. If you know how many hours Polly works, can you determine how much money she earned? Write the corresponding expression.

   b. Use your expression to determine how much Polly earned after working for $3\frac{1}{2}$ hours.

   c. If you know how much money Polly earned, can you determine how long she worked? Write the corresponding expression.

   d. Use your expression to determine how long Polly worked if she earned $52.50.

4. Mitchell delivers newspapers after school, for which he earns $0.09 per paper. Create a table of values that shows the relationship between the number of papers that Mitchell delivers, P, and the amount of money Mitchell earns in dollars, E.

a. If you know how many papers Mitchell delivered, can you determine how much money he earned? Write the corresponding expression.

b. Use your expression to determine how much Mitchell earned by delivering 300 newspapers.

c. If you know how much money Mitchell earned, can you determine how many papers he delivered? Write the corresponding expression.

d. Use your expression to determine how many papers Mitchell delivered if he earned $58.50 last week.

5. Randy is an art dealer who sells reproductions of famous paintings. Copies of the *Mona Lisa* sell for $475.

a. Last year Randy sold $9,975 worth of *Mona Lisa* reproductions. How many did he sell?

b. If Randy wants to increase his sales to at least $15,000 this year, how many copies will he need to sell (without changing the price per painting)?

Lesson 20:   Writing and Evaluating Expressions—Multiplication and Division

© 2019 Great Minds®. eureka-math.org

Mathematical Modeling Exercise

The Italian Villa Restaurant has square tables that the servers can push together to accommodate the customers. Only one chair fits along the side of the square table. Make a model of each situation to determine how many seats will fit around various rectangular tables.

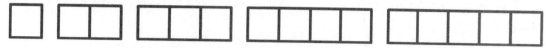

Number of Square Tables	Number of Seats at the Table
1	
2	
3	
4	
5	
50	
200	
T	

Are there any other ways to think about solutions to this problem?

It is impractical to make a model of pushing 50 tables together to make a long rectangle. If we did have a rectangle that long, how many chairs would fit on the long sides of the table?

How many chairs fit on the ends of the long table?

How many chairs fit in all? Record it on your table.

Work with your group to determine how many chairs would fit around a very long rectangular table if 200 square tables were pushed together.

If we let T represent the number of square tables that make one long rectangular table, what is the expression for the number of chairs that will fit around it?

Example

Look at Example 1 with your group. Determine the cost for various numbers of pizzas, and also determine the expression that describes the cost of having P pizzas delivered.

a. Pizza Queen has a special offer on lunch pizzas: $4.00 each. They charge $2.00 to deliver, regardless of how many pizzas are ordered. Determine the cost for various numbers of pizzas, and also determine the expression that describes the cost of having P pizzas delivered.

Number of Pizzas Delivered	Total Cost in Dollars
1	
2	
3	
4	
10	
50	
P	

What mathematical operations did you need to perform to find the total cost?

Lesson 21: Writing and Evaluating Expressions—Multiplication and Addition

© 2019 Great Minds®. eureka-math.org

Suppose our principal wanted to buy a pizza for everyone in our class. Determine how much this would cost.

b. If the booster club had $400 to spend on pizza, what is the greatest number of pizzas they could order?

c. If the pizza price was raised to $5.00 and the delivery price was raised to $3.00, create a table that shows the total cost (pizza plus delivery) of 1, 2, 3, 4, and 5 pizzas. Include the expression that describes the new cost of ordering P pizzas.

Number of Pizzas Delivered	Total Cost in Dollars
1	
2	
3	
4	
5	
P	

Name _____   Date _____

Krystal Klear Cell Phone Company charges $5.00 per month for service. The company also charges $0.10 for each text message sent.

a. Complete the table below to calculate the monthly charges for various numbers of text messages sent.

Number of Text Messages Sent (T)	Total Monthly Bill in Dollars
0	
10	
20	
30	
T	

b. If Suzannah's budget limit is $10 per month, how many text messages can she send in one month?

Name _____   Date _____

Krystal Klear Cell Phone Company charges $5.00 per month for service. The company also charges $0.10 for each text message sent.

    a.   Complete the table below to calculate the monthly charges for various numbers of text messages sent.

Number of Text Messages Sent (T)	Total Monthly Bill in Dollars
0	
10	
20	
30	
T	

    b.   If Suzannah's budget limit is $10 per month, how many text messages can she send in one month?

1. Victoria is purchasing shirts at $8 each for the math team. The company charges $7.25 for shipping and handling, no matter how many shirts are purchased.

   a. Create a table of values that shows the relationship between the number of shirts that Victoria buys, S, and the amount of money Victoria spends, T, in dollars.

   > Like in Lesson 20, I can choose the number of shirts that I use in the table.

Number of Shirts Victoria Buys (S)	Total Cost in Dollars (T)
1	15.25
2	23.25
3	31.25

   b. If you know how many shirts Victoria orders, can you determine how much money she spends? Write the corresponding expression.

   $8S + 7.25$

   > First, I can determine the cost of the shirts, and then I will add on the shipping cost.

   c. Use your expression to determine how much Victoria spent buying 30 shirts.

   $8(30) + 7.25$

   247.25

   > Now I can use the expression I came up with in part (b) to solve when $S = 30$.

   Victoria spent $247.25.

2. When riding in a taxi, Hector pays a $3 flat fee and $2.75 per mile. The relationship between the number of miles, M, and the total cost, C, is shown in the table.

a. Complete the missing values in the table.

Number of Miles (M)	Total Cost in Dollars (C)
1	5.75
2	8.50
3	11.25
4	14
5	**16.75**
6	**19.50**

I can see a pattern in the second column. The cost is increasing by $2.75 each time another mile is added in the first column.

b. Write an expression that shows the cost of taking the taxi for a total of M miles.

$3 + 2.75M$

I add the flat fee to the cost for the miles traveled.

c. If Hector can only spend $47 on the taxi ride, how many miles can he travel?

I can work backwards to figure out how many miles Hector can afford. First, I will subtract the flat fee from the total. Then, I will divide by the price per mile.

$$47 - 3 = 44$$
$$44 \div 2.75 = 16$$

Hector can go 16 miles in the taxi.

Lesson 21:   Writing and Evaluating Expressions—Multiplication and Addition

EUREKA MATH

1. Compact discs (CDs) cost $12 each at the Music Emporium. The company charges $4.50 for shipping and handling, regardless of how many compact discs are purchased.

   a. Create a table of values that shows the relationship between the number of compact discs that Mickey buys, D, and the amount of money Mickey spends, C, in dollars.

Number of CDs Mickey Buys (D)	Total Cost in Dollars (C)
1	
2	
3	

   b. If you know how many CDs Mickey orders, can you determine how much money he spends? Write the corresponding expression.

   c. Use your expression to determine how much Mickey spent buying 8 CDs.

2. Mr. Gee's class orders paperback books from a book club. The books cost $2.95 each. Shipping charges are set at $4.00, regardless of the number of books purchased.

   a. Create a table of values that shows the relationship between the number of books that Mr. Gee's class buys, B, and the amount of money they spend, C, in dollars.

Number of Books Ordered (B)	Amount of Money Spent in Dollars (C)
1	
2	
3	

   b. If you know how many books Mr. Gee's class orders, can you determine how much money they spend? Write the corresponding expression.

   c. Use your expression to determine how much Mr. Gee's class spent buying 24 books.

3. Sarah is saving money to take a trip to Oregon. She received $450 in graduation gifts and saves $120 per week working.

   a. Write an expression that shows how much money Sarah has after working W weeks.

   b. Create a table that shows the relationship between the amount of money Sarah has (M) and the number of weeks she works (W).

Amount of Money Sarah Has (M)	Number of Weeks Worked (W)
	1
	2
	3
	4
	5
	6
	7
	8

   c. The trip will cost $1,200. How many weeks will Sarah have to work to earn enough for the trip?

4. Mr. Gee's language arts class keeps track of how many words per minute are read aloud by each of the students. They collect this oral reading fluency data each month. Below is the data they collected for one student in the first four months of school.

   a. Assume this increase in oral reading fluency continues throughout the rest of the school year. Complete the table to project the reading rate for this student for the rest of the year.

Month	Number of Words Read Aloud in One Minute
September	126
October	131
November	136
December	141
January	
February	
March	
April	
May	
June	

   b. If this increase in oral reading fluency continues throughout the rest of the school year, when would this student achieve the goal of reading 165 words per minute?

   c. The expression for this student's oral reading fluency is $121 + 5m$, where m represents the number of months during the school year. Use this expression to determine how many words per minute the student would read after 12 months of instruction.

     Lesson 21:   Writing and Evaluating Expressions—Multiplication and Addition

EUREKA MATH®

5. When corn seeds germinate, they tend to grow 5 inches in the first week and then 3 inches per week for the remainder of the season. The relationship between the height (H) and the number of weeks since germination (W) is shown below.

    a. Complete the missing values in the table.

Number of Weeks Since Germination (W)	Height of Corn Plant (H)
1	5
2	8
3	11
4	14
5	
6	

    b. The expression for this height is $2 + 3W$. How tall will the corn plant be after 15 weeks of growth?

6. The Honeymoon Charter Fishing Boat Company only allows newlywed couples on their sunrise trips. There is a captain, a first mate, and a deck hand manning the boat on these trips.

    a. Write an expression that shows the number of people on the boat when there are C couples booked for the trip.

    b. If the boat can hold a maximum of 20 people, how many couples can go on the sunrise fishing trip?

Example 1: Folding Paper

Exercises

1. Predict how many times you can fold a piece of paper in half.

    My prediction: _____

2. Before any folding (zero folds), there is only one layer of paper. This is recorded in the first row of the table.
    Fold your paper in half. Record the number of layers of paper that result. Continue as long as possible.

Number of Folds	Number of Paper Layers That Result	Number of Paper Layers Written as a Power of 2
0	1	2^0
1		
2		
3		
4		
5		
6		
7		
8		

a. Are you able to continue folding the paper indefinitely? Why or why not?

b. How could you use a calculator to find the next number in the series?

c. What is the relationship between the number of folds and the number of layers?

d. How is this relationship represented in exponential form of the numerical expression?

e. If you fold a paper f times, write an expression to show the number of paper layers.

3. If the paper were to be cut instead of folded, the height of the stack would double at each successive stage, and it would be possible to continue.

a. Write an expression that describes how many layers of paper result from 16 cuts.

b. Evaluate this expression by writing it in standard form.

Example 2: Bacterial Infection

Bacteria are microscopic single-celled organisms that reproduce in a couple of different ways, one of which is called *binary fission*. In binary fission, a bacterium increases its size until it is large enough to split into two parts that are identical. These two grow until they are both large enough to split into two individual bacteria. This continues as long as growing conditions are favorable.

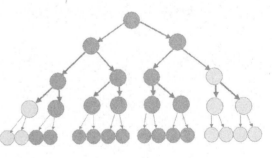

a. Record the number of bacteria that result from each generation.

Generation	Number of Bacteria	Number of Bacteria Written as a Power of 2
1	2	2^1
2	4	2^2
3	8	2^3
4		
5		
6		
7		
8		
9		
10		
11		
12		
13		
14		

b. How many generations would it take until there were over one million bacteria present?

c. Under the right growing conditions, many bacteria can reproduce every 15 minutes. Under these conditions, how long would it take for one bacterium to reproduce itself into more than one million bacteria?

d. Write an expression for how many bacteria would be present after g generations.

Example 3: Volume of a Rectangular Solid

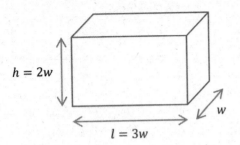

This box has a width, w. The height of the box, h, is twice the width. The length of the box, l, is three times the width. That is, the width, height, and length of a rectangular prism are in the ratio of $1 : 2 : 3$.

For rectangular solids like this, the volume is calculated by multiplying length times width times height.

$$V = l \cdot w \cdot h$$
$$V = 3w \cdot w \cdot 2w$$
$$V = 3 \cdot 2 \cdot w \cdot w \cdot w$$
$$V = 6w^3$$

Follow the above example to calculate the volume of these rectangular solids, given the width, w.

Width in Centimeters (cm)	Volume in Cubic Centimeters (cm^3)
1	
2	
3	
4	
w	

Name _____   Date _____

1. Naomi's allowance is $2.00 per week. If she convinces her parents to double her allowance each week for two months, what will her weekly allowance be at the end of the second month (week 8)?

Week Number	Allowance
1	$2.00
2	
3	
4	
5	
6	
7	
8	
w	

2. Write the expression that describes Naomi's allowance during week w in dollars.

1. Miguel tried a new restaurant on Day 1. On Day 2, he told
    3 friends about the restaurant. On Day 3, each friend told
    3 friends about the restaurant, and the pattern continued
    tripling each day for 9 days.

    > Because the number of people is tripling every time, I can use an exponent to represent the amount of people that hear about the restaurant each day.

    a. Complete the table to show how many people heard
        about the restaurant each day. Write your answers in
        exponential form on the table below.

Day	Number of People	Day	Number of People	Day	Number of People
1	3^0	4	3^3	7	3^6
2	3^1	5	3^4	8	3^7
3	3^2	6	3^5	9	3^8

    b. How many people would be told about the restaurant on Day 9? Represent your answer in
        exponential form and standard form. Use the table above to help solve the problem.

        > To find out how many people will hear about the restaurant on Day 9, I need to solve $3 \times 3 \times 3 \times 3 \times 3 \times 3 \times 3 \times 3$.

    On Day 9, 3^8—or 6, 561—people would hear about the restaurant.

c.   Miguel is estimating that by Day 9 at least 10,000 people will have heard about the restaurant. Is his estimate accurate? Why or why not?

> At least 10,000 means 10,000 or more.

$1 + 3 + 9 + 27 + 81 + 243 + 729 + 2187 + 6561 = 9841$

By Day 9, only 9,841 people will have heard about the restaurant. Although 9,841 is close to 10,000, it is not over 10,000. Miguel estimated that at least 10,000 would have heard about the restaurant, which would mean 10,000 or more.

> To find out how many total people have heard about the restaurant, I need to add the totals from each day.

2.   If an amount of money is invested at an annual interest rate of 9%, it doubles every 8 years. If Van invests $700, how long will it take for his investment to reach $2,800 (assuming he does not contribute any additional funds)?

It will take 16 years to reach $2,800.

> After 8 years, it will reach $1,400. Then after another 8 years, it will double again to $2,800.

1. A checkerboard has 64 squares on it.

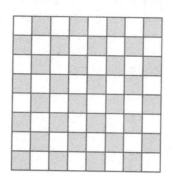

a. If one grain of rice is put on the first square, 2 grains of rice on the second square, 4 grains of rice on the third square, 8 grains of rice on the fourth square, and so on (doubling each time), complete the table to show how many grains of rice are on each square. Write your answers in exponential form on the table below.

Checkerboard Square	Grains of Rice	Checkerboard Square	Grains of Rice	Checkerboard Square	Grains of Rice	Checkerboard Square	Grains of Rice
1		17		33		49	
2		18		34		50	
3		19		35		51	
4		20		36		52	
5		21		37		53	
6		22		38		54	
7		23		39		55	
8		24		40		56	
9		25		41		57	
10		26		42		58	
11		27		43		59	
12		28		44		60	
13		29		45		61	
14		30		46		62	
15		31		47		63	
16		32		48		64	

b. How many grains of rice would be on the last square? Represent your answer in exponential form and standard form. Use the table above to help solve the problem.

c. Would it have been easier to write your answer to part (b) in exponential form or standard form?

2. If an amount of money is invested at an annual interest rate of 6%, it doubles every 12 years. If Alejandra invests $500, how long will it take for her investment to reach $2,000 (assuming she does not contribute any additional funds)?

3. The athletics director at Peter's school has created a phone tree that is used to notify team players in the event a game has to be canceled or rescheduled. The phone tree is initiated when the director calls two captains. During the second stage of the phone tree, the captains each call two players. During the third stage of the phone tree, these players each call two other players. The phone tree continues until all players have been notified. If there are 50 players on the teams, how many stages will it take to notify all of the players?

Opening Exercise

Determine what each symbol stands for, and provide an example.

Symbol	What the Symbol Stands For	Example
=		
>		
<		
≤		
≥		

Example 1

For each equation or inequality your teacher displays, write the equation or inequality, and then substitute 3 for every x. Determine if the equation or inequality results in a true number sentence or a false number sentence.

Exercises

Substitute the indicated value into the variable, and state (in a complete sentence) whether the resulting number sentence is true or false. If true, find a value that would result in a false number sentence. If false, find a value that would result in a true number sentence.

1. $4 + x = 12$. Substitute 8 for x.

2. $3g > 15$. Substitute $4\frac{1}{2}$ for g.

3. $\frac{f}{4} < 2$. Substitute 8 for f.

4. $14.2 \leq h - 10.3$. Substitute 25.8 for h.

5. $4 = \frac{8}{h}$. Substitute 6 for h.

6. $3 > k + \frac{1}{4}$. Substitute $1\frac{1}{2}$ for k.

Lesson 23: True and False Number Sentences

7. $4.5 - d > 2.5$. Substitute 2.5 for d.

8. $8 \geq 32p$. Substitute $\frac{1}{2}$ for p.

9. $\frac{w}{2} < 32$. Substitute 16 for w.

10. $18 \leq 32 - b$. Substitute 14 for b.

Lesson Summary

NUMBER SENTENCE: A *number sentence* is a statement of equality (or inequality) between two numerical expressions.

TRUTH VALUES OF A NUMBER SENTENCE: A number sentence is said to be *true* if both numerical expressions evaluate to the same number; it is said to be *false* otherwise. True and false are called *truth values.*

Number sentences that are inequalities also have truth values. For example, $3 < 4$, $6 + 8 > 15 - 12$, and $(15 + 3)^2 < 1,000 - 32$ are all true number sentences, while the sentence $9 > 3(4)$ is false.

Name _____   Date _____

Substitute the value for the variable, and state in a complete sentence whether the resulting number sentence is true or false. If true, find a value that would result in a false number sentence. If false, find a value that would result in a true number sentence.

1. $15a \geq 75$. Substitute 5 for a.

2. $23 + b = 30$. Substitute 10 for b.

3. $20 > 86 - h$. Substitute 46 for h.

4. $32 \geq 8m$. Substitute 5 for m.

Substitute the value for the variable, and state whether the resulting number sentence is true or false. If true, find a value that would result in a false number sentence. If false, find a value that would result in a true number sentence.

1. $4\frac{1}{2} = 2\frac{3}{4} + g$. Substitute $1\frac{3}{4}$ for g.

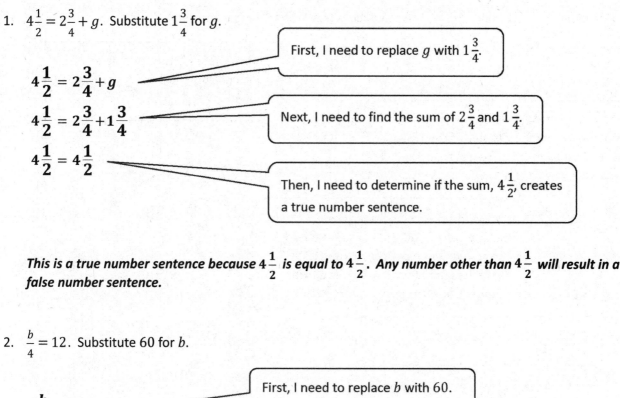

$$4\frac{1}{2} = 2\frac{3}{4} + g$$

First, I need to replace g with $1\frac{3}{4}$.

$$4\frac{1}{2} = 2\frac{3}{4} + 1\frac{3}{4}$$

Next, I need to find the sum of $2\frac{3}{4}$ and $1\frac{3}{4}$.

$$4\frac{1}{2} = 4\frac{1}{2}$$

Then, I need to determine if the sum, $4\frac{1}{2}$, creates a true number sentence.

This is a true number sentence because $4\frac{1}{2}$ is equal to $4\frac{1}{2}$. Any number other than $4\frac{1}{2}$ will result in a false number sentence.

2. $\frac{b}{4} = 12$. Substitute 60 for b.

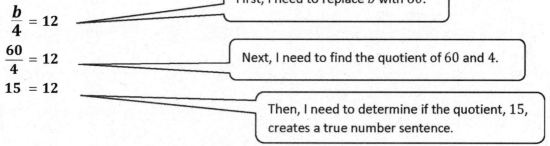

$$\frac{b}{4} = 12$$

First, I need to replace b with 60.

$$\frac{60}{4} = 12$$

Next, I need to find the quotient of 60 and 4.

$$15 = 12$$

Then, I need to determine if the quotient, 15, creates a true number sentence.

This is a false number sentence because when 60 is divided by 4, the quotient is 15. In order for this to be a true number sentence, the quotient must equal 12. To create a true number sentence, the variable, b, must be replaced with a number that, when divided by 4, will create a quotient of 12. When replacing b, the only number that will create a true number sentence is 48.

Create a number sentence using the given variable and symbol. The number sentence you write must be true for the given value of the variable.

3. Variable: m Symbol: $\geq$ The sentence is true when 13 is substituted for m.

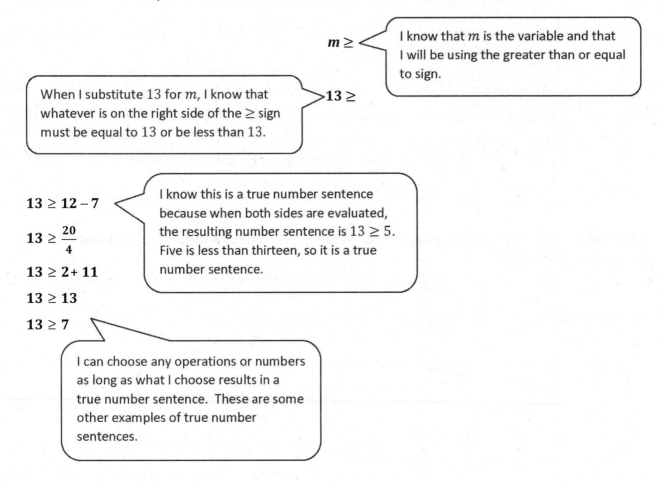

$m \geq$

I know that m is the variable and that I will be using the greater than or equal to sign.

When I substitute 13 for m, I know that whatever is on the right side of the $\geq$ sign must be equal to 13 or be less than 13.

$13 \geq$

$13 \geq 12 - 7$

$13 \geq \dfrac{20}{4}$

$13 \geq 2 + 11$

$13 \geq 13$

$13 \geq 7$

I know this is a true number sentence because when both sides are evaluated, the resulting number sentence is $13 \geq 5$. Five is less than thirteen, so it is a true number sentence.

I can choose any operations or numbers as long as what I choose results in a true number sentence. These are some other examples of true number sentences.

     Lesson 23:   True and False Number Sentences

Substitute the value into the variable, and state (in a complete sentence) whether the resulting number sentence is true or false. If true, find a value that would result in a false number sentence. If false, find a value that would result in a true number sentence.

1. $3\frac{5}{6} = 1\frac{2}{3} + h$. Substitute $2\frac{1}{6}$ for h.

2. $39 > 156g$. Substitute $\frac{1}{4}$ for g.

3. $\frac{f}{4} \le 3$. Substitute 12 for f.

4. $121 - 98 \ge r$. Substitute 23 for r.

5. $\frac{54}{q} = 6$. Substitute 10 for q.

Create a number sentence using the given variable and symbol. The number sentence you write must be true for the given value of the variable.

6. Variable: d Symbol: $\ge$ The sentence is true when 5 is substituted for d.

7. Variable: y Symbol: $\ne$ The sentence is true when 10 is substituted for y.

8. Variable: k Symbol: $<$ The sentence is true when 8 is substituted for k.

9. Variable: a Symbol: $\le$ The sentence is true when 9 is substituted for a.

Opening Exercise

State whether each number sentence is true or false. If the number sentence is false, explain why.

a. $4 + 5 > 9$

b. $3 \cdot 6 = 18$

c. $32 > \dfrac{64}{4}$

d. $78 - 15 < 68$

e. $22 \geq 11 + 12$

Example 1

Write true or false if the number substituted for g results in a true or false number sentence.

Substitute g with	$4g = 32$	$g = 8$	$3g \geq 30$	$g \geq 10$	$\dfrac{g}{2} > 2$	$g > 4$	$30 \geq 38 - g$	$g \geq 8$
8								
4								
2								
0								
10								

Example 2

State when the following equations/inequalities will be true and when they will be false.

a. $r + 15 = 25$

b. $6 - d > 0$

c. $\dfrac{1}{2}f = 15$

d. $\dfrac{y}{3} < 10$

e. $7g \geq 42$

f. $a - 8 \leq 15$

Exercises

Complete the following problems in pairs. State when the following equations and inequalities will be true and when they will be false.

1. $15c > 45$

2. $25 = d - 10$

3. $56 \geq 2e$

4. $\dfrac{h}{5} \geq 12$

5. $45 > h + 29$

6. $4a \leq 16$

7. $3x = 24$

Identify all equality and inequality signs that can be placed into the blank to make a true number sentence.

8. $15 + 9$ _____ 24

9. $8 \cdot 7$ _____ 50

10. $\dfrac{15}{2}$ _____ 10

11. 34 _____ $17 \cdot 2$

12. 18 _____ $24.5 - 6$

Name _____   Date _____

State when the following equations and inequalities will be true and when they will be false.

1. $5g > 45$

2. $14 = 5 + k$

3. $26 - w < 12$

4. $32 \leq a + 8$

5. $2 \cdot h \leq 16$

State when the following equations or inequalities will be true and when they will be false.

1. $72 = 2f$

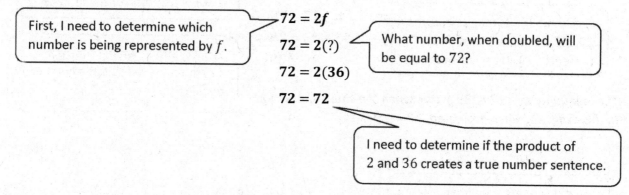

First, I need to determine which number is being represented by f.

$72 = 2f$

$72 = 2(?)$

$72 = 2(36)$

$72 = 72$

What number, when doubled, will be equal to 72?

I need to determine if the product of 2 and 36 creates a true number sentence.

The equation $72 = 2f$ is true only when the value of f is 36 and false when the value of f is any number other than 36. The equation is true when $f = 36$ and false when $f \neq 36$.

2. $m - 12 \leq 29$

$m - 12 \leq 29$

First, I need to think of numbers that are less than or equal to 29.

$m - 12 = 29$

Any number less than or equal to 29 will result in a true number sentence. I'll start with equal.

$41 - 12 = 29$

$29 = 29$

What number, when I subtract 12 from it, will result in a number equal to 29?

The inequality $m - 12 \leq 29$ is true when the value of m is 41. In this example, when evaluated, $29 = 29$.

What if I choose a number greater than 41? Will that also result in a true number sentence?

$42 - 12 \leq 29$

$30 \leq 29$

The inequality $m - 12 \leq 29$ is false when the value of m is greater than 41. In this example, 30 is not equal to 29, nor is it less than 29.

What if I choose a number less than 41? Will that also result in a true number sentence?

$40 - 12 \leq 29$

$28 \leq 29$

The inequality $m - 12 \leq 29$ is true when the value of m is less than 41. In this example, 28 is not equal to 29, but it is less than 29.

Therefore, the inequality $m - 12 \leq 29$ is true when the value of m is less than or equal to 41. It is false when the value of m is more than 41. The inequality is true when $m \leq 41$ and false when $m > 41$.

EUREKA MATH®

State when the following equations and inequalities will be true and when they will be false.

1. $36 = 9k$

2. $67 > f - 15$

3. $\dfrac{v}{9} = 3$

4. $10 + b > 42$

5. $d - 8 \geq 35$

6. $32f < 64$

7. $10 - h \leq 7$

8. $42 + 8 \geq g$

9. $\dfrac{m}{3} = 14$

Division of Fractions—Round 1

Directions: Evaluate each expression and simplify.

Number Correct: _____

1.	9 ones ÷ 3 ones	
2.	$9 \div 3$	
3.	9 tens ÷ 3 tens	
4.	$90 \div 30$	
5.	9 hundreds ÷ 3 hundreds	
6.	$900 \div 300$	
7.	9 halves ÷ 3 halves	
8.	$\frac{9}{2} \div \frac{3}{2}$	
9.	9 fourths ÷ 3 fourths	
10.	$\frac{9}{4} \div \frac{3}{4}$	
11.	$\frac{9}{8} \div \frac{3}{8}$	
12.	$\frac{2}{3} \div \frac{1}{3}$	
13.	$\frac{1}{3} \div \frac{2}{3}$	
14.	$\frac{6}{7} \div \frac{2}{7}$	
15.	$\frac{5}{7} \div \frac{2}{7}$	
16.	$\frac{3}{7} \div \frac{4}{7}$	
17.	$\frac{6}{10} \div \frac{2}{10}$	
18.	$\frac{6}{10} \div \frac{4}{10}$	
19.	$\frac{6}{10} \div \frac{8}{10}$	
20.	$\frac{7}{12} \div \frac{2}{12}$	
21.	$\frac{6}{12} \div \frac{9}{12}$	
22.	$\frac{4}{12} \div \frac{11}{12}$	

23.	$\frac{6}{10} \div \frac{4}{10}$	
24.	$\frac{6}{10} \div \frac{2}{5} = \frac{6}{10} \div \frac{}{10}$	
25.	$\frac{10}{12} \div \frac{5}{12}$	
26.	$\frac{5}{6} \div \frac{5}{12} = \frac{}{12} \div \frac{5}{12}$	
27.	$\frac{10}{12} \div \frac{3}{12}$	
28.	$\frac{10}{12} \div \frac{1}{4} = \frac{10}{12} \div \frac{}{12}$	
29.	$\frac{5}{6} \div \frac{3}{12} = \frac{}{12} \div \frac{3}{12}$	
30.	$\frac{5}{10} \div \frac{2}{10}$	
31.	$\frac{5}{10} \div \frac{1}{5} = \frac{5}{10} \div \frac{}{10}$	
32.	$\frac{1}{2} \div \frac{2}{10} = \frac{}{10} \div \frac{2}{10}$	
33.	$\frac{1}{2} \div \frac{2}{4}$	
34.	$\frac{3}{4} \div \frac{2}{8}$	
35.	$\frac{1}{2} \div \frac{3}{8}$	
36.	$\frac{1}{2} \div \frac{1}{5} = \frac{}{10} \div \frac{}{10}$	
37.	$\frac{2}{4} \div \frac{1}{3}$	
38.	$\frac{1}{4} \div \frac{4}{6}$	
39.	$\frac{3}{4} \div \frac{2}{6}$	
40.	$\frac{5}{6} \div \frac{1}{4}$	
41.	$\frac{2}{9} \div \frac{5}{6}$	
42.	$\frac{5}{9} \div \frac{1}{6}$	
43.	$\frac{1}{2} \div \frac{1}{7}$	
44.	$\frac{5}{7} \div \frac{1}{2}$	

Division of Fractions—Round 2

Number Correct: _____

Improvement: _____

Directions: Evaluate each expression and simplify.

1.	12 ones ÷ 2 ones	
2.	$12 \div 2$	
3.	12 tens ÷ 2 tens	
4.	$120 \div 20$	
5.	12 hundreds ÷ 2 hundreds	
6.	$1{,}200 \div 200$	
7.	12 halves ÷ 2 halves	
8.	$\dfrac{12}{2} \div \dfrac{2}{2}$	
9.	12 fourths ÷ 3 fourths	
10.	$\dfrac{12}{4} \div \dfrac{3}{4}$	
11.	$\dfrac{12}{8} \div \dfrac{3}{8}$	
12.	$\dfrac{2}{4} \div \dfrac{1}{4}$	
13.	$\dfrac{1}{4} \div \dfrac{2}{4}$	
14.	$\dfrac{4}{5} \div \dfrac{2}{5}$	
15.	$\dfrac{2}{5} \div \dfrac{4}{5}$	
16.	$\dfrac{3}{5} \div \dfrac{4}{5}$	
17.	$\dfrac{6}{8} \div \dfrac{2}{8}$	
18.	$\dfrac{6}{8} \div \dfrac{4}{8}$	
19.	$\dfrac{6}{8} \div \dfrac{5}{8}$	
20.	$\dfrac{6}{10} \div \dfrac{2}{10}$	
21.	$\dfrac{7}{10} \div \dfrac{8}{10}$	
22.	$\dfrac{4}{10} \div \dfrac{7}{10}$	

23.	$\dfrac{6}{12} \div \dfrac{4}{12}$	
24.	$\dfrac{6}{12} \div \dfrac{2}{6} = \dfrac{6}{12} \div \dfrac{}{12}$	
25.	$\dfrac{8}{14} \div \dfrac{7}{14}$	
26.	$\dfrac{8}{14} \div \dfrac{1}{2} = \dfrac{8}{14} \div \dfrac{}{14}$	
27.	$\dfrac{11}{14} \div \dfrac{2}{14}$	
28.	$\dfrac{11}{14} \div \dfrac{1}{7} = \dfrac{11}{14} \div \dfrac{}{14}$	
29.	$\dfrac{1}{7} \div \dfrac{6}{14} = \dfrac{}{14} \div \dfrac{6}{14}$	
30.	$\dfrac{7}{18} \div \dfrac{3}{18}$	
31.	$\dfrac{7}{18} \div \dfrac{1}{6} = \dfrac{7}{18} \div \dfrac{}{18}$	
32.	$\dfrac{1}{3} \div \dfrac{12}{18} = \dfrac{}{18} \div \dfrac{12}{18}$	
33.	$\dfrac{1}{6} \div \dfrac{4}{18}$	
34.	$\dfrac{4}{12} \div \dfrac{8}{6}$	
35.	$\dfrac{1}{3} \div \dfrac{3}{15}$	
36.	$\dfrac{2}{6} \div \dfrac{1}{9} = \dfrac{}{18} \div \dfrac{}{18}$	
37.	$\dfrac{1}{6} \div \dfrac{4}{9}$	
38.	$\dfrac{2}{3} \div \dfrac{3}{4}$	
39.	$\dfrac{1}{3} \div \dfrac{3}{5}$	
40.	$\dfrac{1}{7} \div \dfrac{1}{2}$	
41.	$\dfrac{5}{6} \div \dfrac{2}{9}$	
42.	$\dfrac{5}{9} \div \dfrac{2}{6}$	
43.	$\dfrac{5}{6} \div \dfrac{4}{9}$	
44.	$\dfrac{1}{2} \div \dfrac{4}{5}$	

Opening Exercise

Identify a value for the variable that would make each equation or inequality into a true number sentence. Is this the only possible answer? State when the equation or inequality is true using equality and inequality symbols.

a. $3 + g = 15$

b. $30 > 2d$

c. $\dfrac{15}{f} < 5$

d. $42 \leq 50 - m$

Example

Each of the following numbers, if substituted for the variable, makes one of the equations below into a true number sentence. Match the number to that equation: 3, 6, 15, 16, 44.

a. $n + 26 = 32$

b. $n - 12 = 32$

c. $17n = 51$

d. $4^2 = n$

e. $\dfrac{n}{3} = 5$

Lesson 25:   Finding Solutions to Make Equations True

Lesson Summary

VARIABLE: A *variable* is a symbol (such as a letter) that is a placeholder for a number.

A variable is a placeholder for "a number" that does not "vary."

EXPRESSION: An *expression* is a numerical expression, or it is the result of replacing some (or all) of the numbers in a numerical expression with variables.

EQUATION: An *equation* is a statement of equality between two expressions.

If A and B are two expressions in the variable x, then $A = B$ is an equation in the variable x.

Name _____ Date _____

Find the solution to each equation.

1. $7f = 49$

2. $1 = \dfrac{r}{12}$

3. $1.5 = d + 0.8$

4. $9^2 = h$

5. $q = 45 - 19$

6. $40 = \dfrac{1}{2}p$

Solutions to Equations

When solving equations, there is only one number that the variable can represent that will result in a true number sentence.

Find the solution to each equation.

1. $5^3 = b$

> Each side of the equation must evaluate to the same number.

> 5^3 is evaluated by multiplying the base, 5, by itself the number of times of the exponent, 3. $5 \times 5 \times 5 = 125$.

$$5^3 = b$$
$$125 = b$$
$$125 = 125$$

> Since 5^3 is 125, the value of b is the same since they are equal. b must equal 125.

2. $6n = 72$

> What number, when multiplied by 6, will result in the product, 72?

$$6n = 72$$
$$6 \times ? = 72$$
$$6 \times 12 = 72$$
$$72 = 72$$

> The right side of the equation is 72. Because this is an equation, the product of 6 and the number n represents must also equal 72.

> The value of n is 12. It is the only number that can replace n to result in a true number sentence.

EUREKA MATH®

3. $\dfrac{36}{h} = 4$

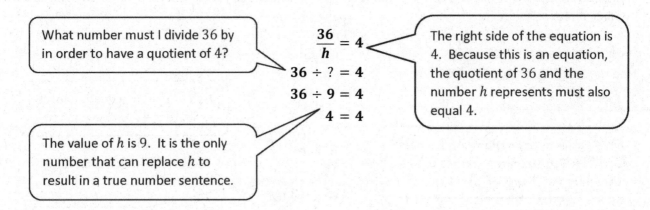

What number must I divide 36 by in order to have a quotient of 4?

$$\dfrac{36}{h} = 4$$

$36 \div ? = 4$

$36 \div 9 = 4$

$4 = 4$

The right side of the equation is 4. Because this is an equation, the quotient of 36 and the number h represents must also equal 4.

The value of h is 9. It is the only number that can replace h to result in a true number sentence.

Lesson 25:   Finding Solutions to Make Equations True

Find the solution to each equation.

1. $4^3 = y$

2. $8a = 24$

3. $32 = g - 4$

4. $56 = j + 29$

5. $\dfrac{48}{r} = 12$

6. $k = 15 - 9$

7. $x \cdot \dfrac{1}{5} = 60$

8. $m + 3.45 = 12.8$

9. $a = 1^5$

Exercise 1

Solve each equation. Use both tape diagrams and algebraic methods for each problem. Use substitution to check your answers.

a. $b + 9 = 15$

b. $12 = 8 + c$

Lesson 26:   One-Step Equations—Addition and Subtraction

Exercise 2

Given the equation $d - 5 = 7$:

    a.   Demonstrate how to solve the equation using tape diagrams.

    b.   Demonstrate how to solve the equation algebraically.

    c.   Check your answer.

    Lesson 26:   One-Step Equations—Addition and Subtraction

Exercise 3

Solve each problem, and show your work. You may choose which method (tape diagrams or algebraically) you prefer. Check your answers after solving each problem.

a. $e + 12 = 20$

b. $f - 10 = 15$

c. $g - 8 = 9$

Name _____   Date _____

1. If you know the answer, state it. Then, use a tape diagram to demonstrate why this is the correct answer. If you do not know the answer, find the solution using a tape diagram.

$$j + 12 = 25$$

2. Find the solution to the equation algebraically. Check your answer.

$$k - 16 = 4$$

Find the solution to the equation using tape diagrams.

1. $x - 5 = 15$

> This equation is stating that when I subtract 5 from a number (in this case the number is represented by x), then the result is 15. What must that number be that is being represented by x? There is only one number that can make this equation true. I will start with x.

> This tape diagram represents the number that will replace x. It represents the number I am taking 5 from in order to find the difference of 15.

> This tape diagram shows that when I take 5 away from the number that is being represented by x, the result is 15.

x

x

5   15

$x - 5$

$x - 5 + 5$

> There are some things I notice here. The remaining 15 is equal to the quantity $x - 5$. This is stated in the problem. When I combine the 5 and the $x - 5$ in the diagram, it is equal to x because $x - 5 + 5 = x$. This is also supported by the knowledge of the properties of operations I learned in Lessons 1–4.

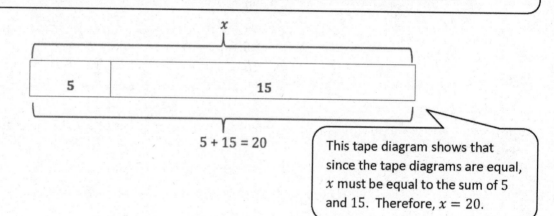

x

5   15

$5 + 15 = 20$

> This tape diagram shows that since the tape diagrams are equal, x must be equal to the sum of 5 and 15. Therefore, $x = 20$.

 EUREKA MATH®

Lesson 26:   One-Step Equations—Addition and Subtraction

303

Find the solution to the equation algebraically. Check your answer.

2. $x - 5 = 15$

$$x - 5 = 15$$
$$x - 5 + 5 = 15 + 5$$
$$x = 15 + 5$$
$$x = 20$$

> $x - 5 + 5 = x$. This is supported by the identity that states that if you take a number away from another number, then add it back in, the result is the first number you began with.

Substitution is a common way to check solutions to equations. Substitute the value for x back into the equation, and evaluate to see if a true number sentence results.

$$x - 5 = 15$$
$$20 - 5 = 15$$
$$15 = 15$$

This is a true number sentence, so the solution, $x = 20$, is correct.

Many times, students confuse the check with the correct solution since they are often different numbers. In order to avoid this problem, students are encouraged to substitute the solution back into the identity, where they will find that the solution and the check will be the same number, resulting in less confusion.

$$x - 5 = 15$$
$$x - 5 + 5 = 15 + 5$$
$$20 - 5 + 5 = 15 + 5$$
$$20 = 20$$

This method also results in a true number sentence and shows that x must equal 20 in order for the equation to be true.

Find the solution to the equation using tape diagrams.

3. $y + 7 = 17$

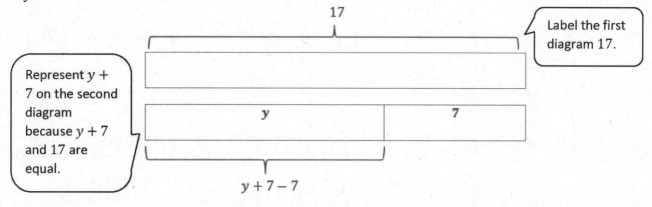

> Label the first diagram 17.

> Represent $y + 7$ on the second diagram because $y + 7$ and 17 are equal.

To find the value that y is representing, subtract 7 from it. In order to do that, 7 must also be subtracted from 17 because $y - 7$ and 17 are equal.

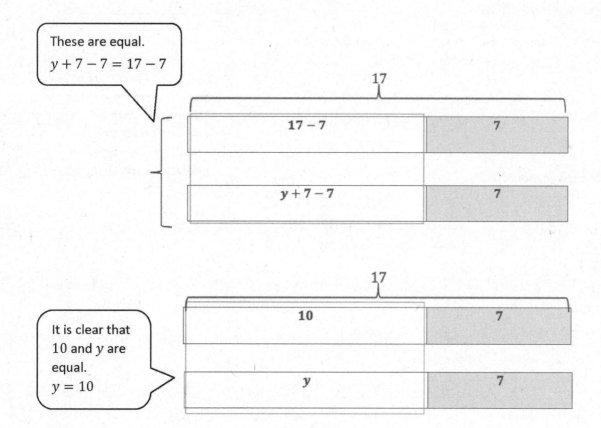

> These are equal.
> $y + 7 - 7 = 17 - 7$

> It is clear that 10 and y are equal.
> $y = 10$

Find the solution to the equation algebraically. Check your answer.

4. $y + 7 = 17$

$$y + 7 = 17$$
$$y + 7 - 7 = 17 - 7$$
$$y = 17 - 7$$
$$y = 10$$

To check work, substitute y with the solution 10 back into the identity to see if it results in a true number sentence.

$$y + 7 = 17$$
$$10 + 7 = 17$$
$$10 + 7 - 7 = 17 - 7$$
$$10 = 10$$

This is a true number sentence, so the solution, $y = 10$, is correct.

Because the identity $w + x - x = w$, I know that I can subtract 7 from $y + 7$ to determine what y represents. Because $y + 7$ and 17 are equal, I also need to subtract 7 from 17.

Identify the mistake in the problem below. Then, correct the mistake.

5. $r + 12 = 32$

$$r + 12 = 32$$
$$r + 12 - 12 = 32 + 12$$
$$r = 44$$

The mistake here is adding 12 on the right side of the equation instead of subtracting 12.

The correct answer should be:

$$r + 12 = 32$$
$$r + 12 - 12 = 32 - 12$$
$$r = 20$$

Lesson 26:   One-Step Equations—Addition and Subtraction

© 2019 Great Minds®. eureka-math.org

1. Find the solution to the equation below using tape diagrams. Check your answer.

$$m - 7 = 17$$

2. Find the solution of the equation below algebraically. Check your answer.

$$n + 14 = 25$$

3. Find the solution of the equation below using tape diagrams. Check your answer.

$$p + 8 = 18$$

4. Find the solution to the equation algebraically. Check your answer.

$$g - 62 = 14$$

5. Find the solution to the equation using the method of your choice. Check your answer.

$$m + 108 = 243$$

6. Identify the mistake in the problem below. Then, correct the mistake.

$$p - 21 = 34$$
$$p - 21 - 21 = 34 - 21$$
$$p = 13$$

7. Identify the mistake in the problem below. Then, correct the mistake.

$$q + 18 = 22$$
$$q + 18 - 18 = 22 + 18$$
$$q = 40$$

8. Match the equation with the correct solution on the right.

$r + 10 = 22$ $r = 10$

$r - 15 = 5$ $r = 20$

$r - 18 = 14$ $r = 12$

$r + 5 = 15$ $r = 32$

Example 1

Solve $3z = 9$ using tape diagrams and algebraically. Then, check your answer.

First, draw two tape diagrams, one to represent each side of the equation.

If 9 had to be split into three groups, how big would each group be?

Demonstrate the value of z using tape diagrams.

How can we demonstrate this algebraically?

How does this get us the value of z?

How can we check our answer?

Solve $\dfrac{y}{4} = 2$ using tape diagrams and algebraically. Then, check your answer.

First, draw two tape diagrams, one to represent each side of the equation.

If the first tape diagram shows the size of $y \div 4$, how can we draw a tape diagram to represent y?

Draw this tape diagram.

What value does each $y \div 4$ section represent? How do you know?

How can you use a tape diagram to show the value of y?

How can we demonstrate this algebraically?

How does this help us find the value of y?

How can we check our answer?

Exercises

1. Use tape diagrams to solve the following problem: $3m = 21$.

2. Solve the following problem algebraically: $15 = \dfrac{n}{5}$.

3. Calculate the solution of the equation using the method of your choice: $4p = 36$.

4. Examine the tape diagram below, and write an equation it represents. Then, calculate the solution to the equation using the method of your choice.

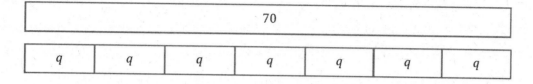

5. Write a multiplication equation that has a solution of 12. Use tape diagrams to prove that your equation has a solution of 12.

6. Write a division equation that has a solution of 12. Prove that your equation has a solution of 12 using algebraic methods.

Name _____ Date _____

Calculate the solution to each equation below using the indicated method. Remember to check your answers.

1. Use tape diagrams to find the solution of $\dfrac{r}{10} = 4$.

2. Find the solution of $64 = 16u$ algebraically.

3. Use the method of your choice to find the solution of $12 = 3v$.

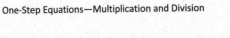

Find the solution to the equation using tape diagrams.

1. $24 = 6x$

This equation is stating that when I multiply a number (in this case it is represented by x) and 6, the product is 24. What must that number be that is being represented by x? There is only one number that can make this equation true. I'm going to start with 24.

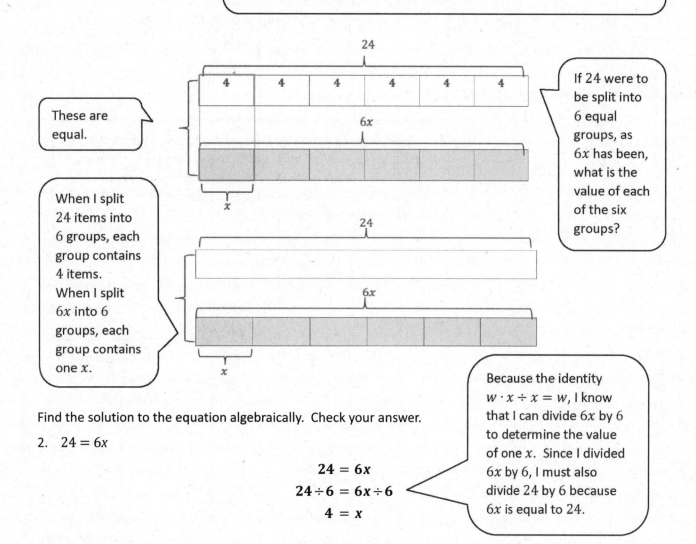

These are equal.

If 24 were to be split into 6 equal groups, as $6x$ has been, what is the value of each of the six groups?

When I split 24 items into 6 groups, each group contains 4 items.
When I split $6x$ into 6 groups, each group contains one x.

Find the solution to the equation algebraically. Check your answer.

2. $24 = 6x$

$$24 = 6x$$
$$24 \div 6 = 6x \div 6$$
$$4 = x$$

Because the identity $w \cdot x \div x = w$, I know that I can divide $6x$ by 6 to determine the value of one x. Since I divided $6x$ by 6, I must also divide 24 by 6 because $6x$ is equal to 24.

Substitute the solution back into the equation, and determine if the result is a true number sentence.

$$24 = 6x$$
$$24 = 6 \cdot 4$$
$$24 \div 6 = 6 \cdot 4 \div 6$$
$$4 = 4$$

This results in a true number sentence and shows that x must equal 4 in order for the equation to be true.

Find the solution to the equation using tape diagrams.

3. $\frac{y}{3} = 21$

This equation is stating that when I divide a number (in this case it is represented by y) and 3, the quotient is 21. What must that number be that is being represented by y? There is only one number that can make this equation true. I'm going to start with $y \div 3$.

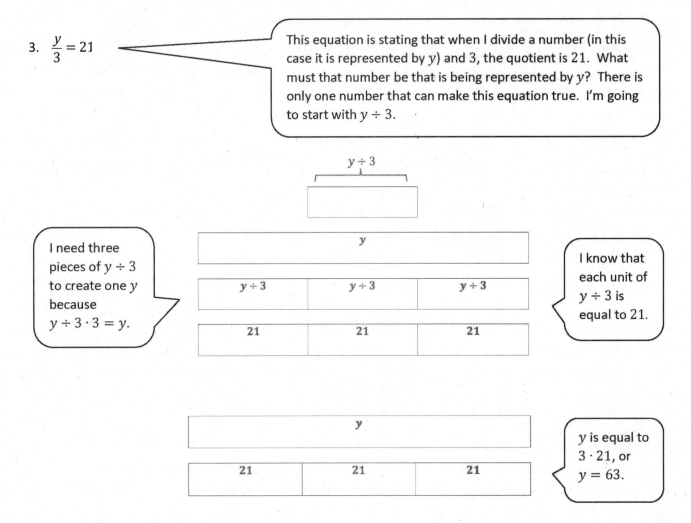

I need three pieces of $y \div 3$ to create one y because $y \div 3 \cdot 3 = y$.

I know that each unit of $y \div 3$ is equal to 21.

y is equal to $3 \cdot 21$, or $y = 63$.

Find the solution to the equation algebraically. Check your answer.

4. $\dfrac{y}{3} = 21$

$$\frac{y}{3} = 21$$

$$y \div 3 = 21$$

$$y \div 3 \cdot 3 = 21 \cdot 3$$

$$y = 63$$

Because the identity $w \div x \cdot x = w$, I know that I can multiply $\dfrac{y}{3}$ by 3 to determine the value of one y. Since I multiplied $\dfrac{y}{3}$ by 3, I must also multiply 21 by 3 because $\dfrac{y}{3}$ is equal to 21.

$$\frac{y}{3} = 21$$

$$63 \div 3 = 21$$

$$63 \div 3 \cdot 3 = 21 \cdot 3$$

$$63 = 63$$

1. Use tape diagrams to calculate the solution of $30 = 5w$. Then, check your answer.

2. Solve $12 = \dfrac{x}{4}$ algebraically. Then, check your answer.

3. Use tape diagrams to calculate the solution of $\dfrac{y}{5} = 15$. Then, check your answer.

4. Solve $18z = 72$ algebraically. Then, check your answer.

5. Write a division equation that has a solution of 8. Prove that your solution is correct by using tape diagrams.

6. Write a multiplication equation that has a solution of 8. Solve the equation algebraically to prove that your solution is correct.

7. When solving equations algebraically, Meghan and Meredith each got a different solution. Who is correct? Why did the other person not get the correct answer?

Meghan	Meredith
$\dfrac{y}{2} = 4$	$\dfrac{y}{2} = 4$
$\dfrac{y}{2} \cdot 2 = 4 \cdot 2$	$\dfrac{y}{2} \div 2 = 4 \div 2$
$y = 8$	$y = 2$

Number Correct: _____

Addition of Decimals II—Round 1

Directions: Evaluate each expression.

1.	2.5 + 4	
2.	2.5 + 0.4	
3.	2.5 + 0.04	
4.	2.5 + 0.004	
5.	2.5 + 0.0004	
6.	6 + 1.3	
7.	0.6 + 1.3	
8.	0.06 + 1.3	
9.	0.006 + 1.3	
10.	0.0006 + 1.3	
11.	0.6 + 13	
12.	7 + 0.2	
13.	0.7 + 0.02	
14.	0.07 + 0.2	
15.	0.7 + 2	
16.	7 + 0.02	
17.	6 + 0.3	
18.	0.6 + 0.03	
19.	0.06 + 0.3	
20.	0.6 + 3	
21.	6 + 0.03	
22.	0.6 + 0.3	

23.	4.5 + 3.1	
24.	4.5 + 0.31	
25.	4.5 + 0.031	
26.	0.45 + 0.031	
27.	0.045 + 0.031	
28.	12 + 0.36	
29.	1.2 + 3.6	
30.	1.2 + 0.36	
31.	1.2 + 0.036	
32.	0.12 + 0.036	
33.	0.012 + 0.036	
34.	0.7 + 3	
35.	0.7 + 0.3	
36.	0.07 + 0.03	
37.	0.007 + 0.003	
38.	5 + 0.5	
39.	0.5 + 0.5	
40.	0.05 + 0.05	
41.	0.005 + 0.005	
42.	0.11 + 19	
43.	1.1 + 1.9	
44.	0.11 + 0.19	

Number Correct: _____

Improvement: _____

Addition of Decimals II—Round 2

Directions: Evaluate each expression.

1.	$7.4 + 3$	
2.	$7.4 + 0.3$	
3.	$7.4 + 0.03$	
4.	$7.4 + 0.003$	
5.	$7.4 + 0.0003$	
6.	$6 + 2.2$	
7.	$0.6 + 2.2$	
8.	$0.06 + 2.2$	
9.	$0.006 + 2.2$	
10.	$0.0006 + 2.2$	
11.	$0.6 + 22$	
12.	$7 + 0.8$	
13.	$0.7 + 0.08$	
14.	$0.07 + 0.8$	
15.	$0.7 + 8$	
16.	$7 + 0.08$	
17.	$5 + 0.4$	
18.	$0.5 + 0.04$	
19.	$0.05 + 0.4$	
20.	$0.5 + 4$	
21.	$5 + 0.04$	
22.	$5 + 0.4$	

23.	$3.6 + 2.3$	
24.	$3.6 + 0.23$	
25.	$3.6 + 0.023$	
26.	$0.36 + 0.023$	
27.	$0.036 + 0.023$	
28.	$0.13 + 56$	
29.	$1.3 + 5.6$	
30.	$1.3 + 0.56$	
31.	$1.3 + 0.056$	
32.	$0.13 + 0.056$	
33.	$0.013 + 0.056$	
34.	$2 + 0.8$	
35.	$0.2 + 0.8$	
36.	$0.02 + 0.08$	
37.	$0.002 + 0.008$	
38.	$0.16 + 14$	
39.	$1.6 + 1.4$	
40.	$0.16 + 0.14$	
41.	$0.016 + 0.014$	
42.	$15 + 0.15$	
43.	$1.5 + 1.5$	
44.	$0.15 + 0.15$	

Mathematical Modeling Exercise

Juan has gained 20 lb. since last year. He now weighs 120 lb. Rashod is 15 lb. heavier than Diego. Rashod weighs the same amount this year that Juan weighed last year. How much does Diego weigh? Let j represent Juan's weight last year in pounds, and let d represent Diego's weight in pounds.

Draw a tape diagram to represent Juan's weight.

Draw a tape diagram to represent Rashod's weight.

Draw a tape diagram to represent Diego's weight.

What would combining all three tape diagrams look like?

Write an equation to represent Juan's tape diagram.

Write an equation to represent Rashod's tape diagram.

How can we use the final tape diagram or the equations above to answer the question presented?

Calculate Diego's weight.

We can use identities to defend our thought that $d + 35 - 35 = d$.

Does your answer make sense?

Lesson 28:   Two-Step Problems—All Operations

© 2019 Great Minds®. eureka-math.org

Example 1

Marissa has twice as much money as Frank. Christina has $20 more than Marissa. If Christina has $100, how much money does Frank have? Let f represent the amount of money Frank has in dollars and m represent the amount of money Marissa has in dollars.

Draw a tape diagram to represent the amount of money Frank has.

Draw a tape diagram to represent the amount of money Marissa has.

Draw a tape diagram to represent the amount of money Christina has.

Which tape diagram provides enough information to determine the value of the variable m?

Write and solve the equation.

The identities we have discussed throughout the module solidify that $m + 20 - 20 = m$.

What does the 80 represent?

Now that we know Marissa has $80, how can we use this information to find out how much money Frank has?

Write an equation.

Solve the equation.

Once again, the identities we have used throughout the module can solidify that $2f \div 2 = f$.

What does the 40 represent?

Does 40 make sense in the problem?

Exercises

<u>**Station One:**</u> **Use tape diagrams to solve the problem.**

Raeana is twice as old as Madeline, and Laura is 10 years older than Raeana. If Laura is 50 years old, how old is Madeline? Let *m* represent Madeline's age in years, and let *r* represent Raeana's age in years.

Station Two: **Use tape diagrams to solve the problem.**

Carli has 90 apps on her phone. Braylen has half the amount of apps as Theiss. If Carli has three times the amount of apps as Theiss, how many apps does Braylen have? Let b represent the number of Braylen's apps and t represent the number of Theiss's apps.

<u>Station Three</u>: Use tape diagrams to solve the problem.

Reggie ran for 180 yards during the last football game, which is 40 more yards than his previous personal best. Monte ran 50 more yards than Adrian during the same game. If Monte ran the same amount of yards Reggie ran in one game for his previous personal best, how many yards did Adrian run? Let r represent the number of yards Reggie ran during his previous personal best and a represent the number of yards Adrian ran.

Station Four: Use tape diagrams to solve the problem.

Lance rides his bike downhill at a pace of 60 miles per hour. When Lance is riding uphill, he rides 8 miles per hour slower than on flat roads. If Lance's downhill speed is 4 times faster than his flat-road speed, how fast does he travel uphill?

Let f represent Lance's pace on flat roads in miles per hour and u represent Lance's pace uphill in miles per hour.

EUREKA MATH®

Name _____ Date _____

Use tape diagrams and equations to solve the problem with visual models and algebraic methods.

Alyssa is twice as old as Brittany, and Jazmyn is 15 years older than Alyssa. If Jazmyn is 35 years old, how old is Brittany? Let a represent Alyssa's age in years and b represent Brittany's age in years.

Use tape diagrams to solve each problem. Then create a set of equations to solve algebraically.

1. Malcolm scored 18 points in tonight's football game, which is 6 points more than his personal best. Jamal scored 2 more points than Alan in tonight's game. Jamal scored the same number of points as Malcolm's personal best. Let m represent the number of points Malcolm scored during his personal best and a represent the number of points Alan scored during tonight's game.

   a. How many points did Alan score during the game?

   Alan scored 10 points during tonight's game.

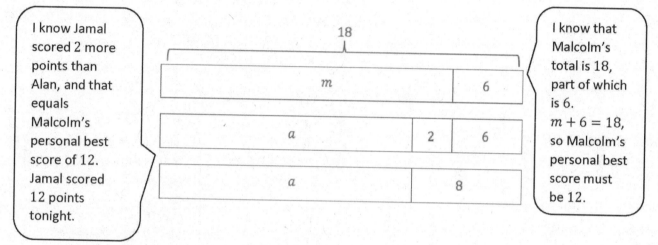

I know Jamal scored 2 more points than Alan, and that equals Malcolm's personal best score of 12. Jamal scored 12 points tonight.

I know that Malcolm's total is 18, part of which is 6. $m + 6 = 18$, so Malcolm's personal best score must be 12.

   Equation for Malcolm's Tape Diagram: $m + 6 = 18$

   Equation for Jamal's Tape Diagram:

$$a + 2 + 6 = 18$$
$$a + 8 = 18$$
$$a + 8 - 8 = 18 - 8$$
$$a = 10$$

Alan's score, plus the 2 more points Jamal scored, plus the 6 extra points that Malcolm scored over his personal best equals Malcolm's total points from tonight's game, 18.

   b. What was the total number of points these three boys scored at the end of tonight's game?

   Malcolm's points + Jamal's points + Alan's points: $18 + 12 + 10 = 40$. ***The total number of points the three boys scored at the end of the game was*** 40.

2. The type of customers at the local bank varies throughout the day. During Saturday's hours, the bank manager collected data to see why the customers were coming into the bank on a Saturday. There were 140 people who made deposits. There were 15 more customers who checked their balances than there were who made withdrawals. The number of customers who made deposits was 5 times as many as those who came in to check their balances. How many customers made withdrawals? How many customers checked their balances? Let w represent the number of customers who made withdrawals, and let b represent the number of customers who only checked their balances.

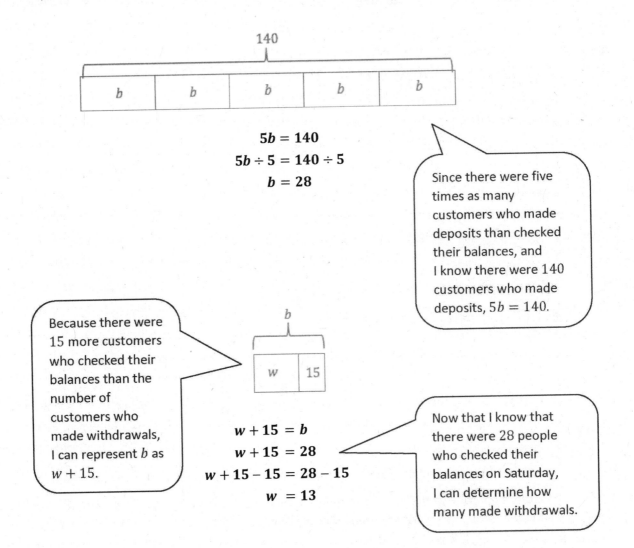

$$5b = 140$$
$$5b \div 5 = 140 \div 5$$
$$b = 28$$

Since there were five times as many customers who made deposits than checked their balances, and I know there were 140 customers who made deposits, $5b = 140$.

Because there were 15 more customers who checked their balances than the number of customers who made withdrawals, I can represent b as $w + 15$.

$$w + 15 = b$$
$$w + 15 = 28$$
$$w + 15 - 15 = 28 - 15$$
$$w = 13$$

Now that I know that there were 28 people who checked their balances on Saturday, I can determine how many made withdrawals.

13 customers made withdrawals on Saturday, and 28 customers checked their balances.

Use tape diagrams to solve each problem.

1. Dwayne scored 55 points in the last basketball game, which is 10 points more than his previous personal best. Lebron scored 15 points more than Chris in the same game. Lebron scored the same number of points as Dwayne's previous personal best. Let d represent the number of points Dwayne scored during his previous personal best and c represent the number of Chris's points.

    a. How many points did Chris score during the game?

    b. If these are the only three players who scored, what was the team's total number of points at the end of the game?

2. The number of customers at Yummy Smoothies varies throughout the day. During the lunch rush on Saturday, there were 120 customers at Yummy Smoothies. The number of customers at Yummy Smoothies during dinner time was 10 customers fewer than the number during breakfast. The number of customers at Yummy Smoothies during lunch was 3 times more than during breakfast. How many people were at Yummy Smoothies during breakfast? How many people were at Yummy Smoothies during dinner? Let d represent the number of customers at Yummy Smoothies during dinner and b represent the number of customers at Yummy Smoothies during breakfast.

3. Karter has 24 T-shirts. Karter has 8 fewer pairs of shoes than pairs of pants. If the number of T-shirts Karter has is double the number of pants he has, how many pairs of shoes does Karter have? Let p represent the number of pants Karter has and s represent the number of pairs of shoes he has.

4. Darnell completed 35 push-ups in one minute, which is 8 more than his previous personal best. Mia completed 6 more push-ups than Katie. If Mia completed the same amount of push-ups as Darnell completed during his previous personal best, how many push-ups did Katie complete? Let d represent the number of push-ups Darnell completed during his previous personal best and k represent the number of push-ups Katie completed.

5. Justine swims freestyle at a pace of 150 laps per hour. Justine swims breaststroke 20 laps per hour slower than she swims butterfly. If Justine's freestyle speed is three times faster than her butterfly speed, how fast does she swim breaststroke? Let b represent Justine's butterfly speed in laps per hour and r represent Justine's breaststroke speed in laps per hour.

Example

The school librarian, Mr. Marker, knows the library has 1,400 books but wants to reorganize how the books are displayed on the shelves. Mr. Marker needs to know how many fiction, nonfiction, and resource books are in the library. He knows that the library has four times as many fiction books as resource books and half as many nonfiction books as fiction books. If these are the only types of books in the library, how many of each type of book are in the library?

Draw a tape diagram to represent the total number of books in the library.

Draw two more tape diagrams, one to represent the number of fiction books in the library and one to represent the number of resource books in the library.

- Resource Books:

- Fiction Books:

What variable should we use throughout the problem?

Write the relationship between resource books and fiction books algebraically.

Draw a tape diagram to represent the number of nonfiction books.

How did you decide how many sections this tape diagram would have?

Represent the number of nonfiction books in the library algebraically.

Use the tape diagrams we drew to solve the problem.

Write an equation that represents the tape diagram.

Determine the value of r.

Lesson 29:   Multi-Step Problems—All Operations

How many fiction books are in the library?

How many nonfiction books are in the library?

Set up a table with four columns, and label each column.

How many fiction books are in the library?

How many nonfiction books are in the library?

How many resource books are in the library?

Does the library have four times as many fiction books as resource books?

Does the library have half as many nonfiction books as fiction books?

Does the library have 1,400 books?

Exercises

Solve each problem below using tables and algebraic methods. Then, check your answers with the word problems.

1. Indiana Ridge Middle School wanted to add a new school sport, so they surveyed the students to determine which sport is most popular. Students were able to choose among soccer, football, lacrosse, or swimming. The same number of students chose lacrosse and swimming. The number of students who chose soccer was double the number of students who chose lacrosse. The number of students who chose football was triple the number of students who chose swimming. If 434 students completed the survey, how many students chose each sport?

2. At Prairie Elementary School, students are asked to pick their lunch ahead of time so the kitchen staff will know what to prepare. On Monday, 6 times as many students chose hamburgers as chose salads. The number of students who chose lasagna was one third the number of students who chose hamburgers. If 225 students ordered lunch, how many students chose each option if hamburger, salad, and lasagna were the only three options?

3. The art teacher, Mr. Gonzalez, is preparing for a project. In order for students to have the correct supplies,
    Mr. Gonzalez needs 10 times more markers than pieces of construction paper. He needs the same number of
    bottles of glue as pieces of construction paper. The number of scissors required for the project is half the number
    of pieces of construction paper. If Mr. Gonzalez collected 400 items for the project, how many of each supply did he
    collect?

4. The math teacher, Ms. Zentz, is buying appropriate math tools to use throughout the year. She is planning on
    buying twice as many rulers as protractors. The number of calculators Ms. Zentz is planning on buying is one
    quarter of the number of protractors. If Ms. Zentz buys 65 items, how many protractors does Ms. Zentz buy?

Name _____   Date _____

Solve the problem using tables and equations, and then check your answer with the word problem. Try to find the answer only using two rows of numbers on your table.

A pet store owner, Byron, needs to determine how much food he needs to feed the animals. Byron knows that he needs to order the same amount of bird food as hamster food. He needs four times as much dog food as bird food and needs half the amount of cat food as dog food. If Byron orders 600 packages of animal food, how much dog food does he buy? Let b represent the number of packages of bird food Byron purchased for the pet store.

Multistep Problems

Solve the problem using a table, and then check your answer with the word problem.

Camille uses four times as many cups of broth as she does cups of milk in a recipe and double the amount of flour as milk.

a. If Camille uses 14 cups of these ingredients in the recipe, how many of each does she use?

> Begin with 1 cup of milk.

> To find out how many cups of flour, double the amount of milk.

> Since there are four times as many cups of broth than milk, multiply 1 cup of milk by four.

> If she uses 14 cups of ingredients, that is twice as many as the original total of 7 cups. If we double 7 cups, we need to double the rest of the original number of cups.

Number of Cups of Milk	Number of Cups of Flour	Number of Cups of Broth	Total Number of Cups Used
1	2	4	7
2	4	8	14

Camille would use 2 cups of milk, 4 cups of flour, and 8 cups of broth. This makes sense because 8 cups of broth is four times as many as 2 cups of milk, and 4 cups of flour is double 2 cups of milk.

b. Support your answer with equations.

Let x represent the number of cups of milk used in the recipe. 2x represents the number of cups of flour, and 4x represents the number of cups of broth. Added together, they equal 14 cups.

$$x + 2x + 4x = 14$$
$$7x = 14$$
$$7x \div 7 = 14 \div 7$$
$$x = 2$$

If $x = 2$, then $2x = 2(2) = 4$, and $4x = 4(2) = 8$.

Create tables to solve the problems, and then check your answers with the word problems.

1. On average, a baby uses three times the number of large diapers as small diapers and double the number of medium diapers as small diapers.

    a. If the average baby uses 2,940 diapers, sizes small, medium, and large, how many of each size would be used?

    b. Support your answer with equations.

2. Tom has three times as many pencils as pens but has a total of 100 writing utensils.

    a. How many pencils does Tom have?

    b. How many more pencils than pens does Tom have?

3. Serena's mom is planning her birthday party. She bought balloons, plates, and cups. Serena's mom bought twice as many plates as cups. The number of balloons Serena's mom bought was half the number of cups.

    a. If Serena's mom bought 84 items, how many of each item did she buy?

    b. Tammy brought 12 balloons to the party. How many total balloons were at Serena's birthday party?

    c. If half the plates and all but four cups were used during the party, how many plates and cups were used?

4. Elizabeth has a lot of jewelry. She has four times as many earrings as watches but half the number of necklaces as earrings. Elizabeth has the same number of necklaces as bracelets.

    a. If Elizabeth has 117 pieces of jewelry, how many earrings does she have?

    b. Support your answer with an equation.

5. Claudia was cooking breakfast for her entire family. She made double the amount of chocolate chip pancakes as she did regular pancakes. She only made half as many blueberry pancakes as she did regular pancakes. Claudia also knows her family loves sausage, so she made triple the amount of sausage as blueberry pancakes.

    a. How many of each breakfast item did Claudia make if she cooked 90 items in total?

    b. After everyone ate breakfast, there were 4 chocolate chip pancakes, 5 regular pancakes, 1 blueberry pancake, and no sausage left. How many of each item did the family eat?

6. During a basketball game, Jeremy scored triple the number of points as Donovan. Kolby scored double the number of points as Donovan.

    a. If the three boys scored 36 points, how many points did each boy score?

    b. Support your answer with an equation.

Number Correct: _____

Subtraction of Decimals—Round 1

Directions: Evaluate each expression.

1.	55 − 50	
2.	55 − 5	
3.	5.5 − 5	
4.	5.5 − 0.5	
5.	88 − 80	
6.	88 − 8	
7.	8.8 − 8	
8.	8.8 − 0.8	
9.	33 − 30	
10.	33 − 3	
11.	3.3 − 3	
12.	1 − 0.3	
13.	1 − 0.03	
14.	1 − 0.003	
15.	0.1 − 0.03	
16.	4 − 0.8	
17.	4 − 0.08	
18.	4 − 0.008	
19.	0.4 − 0.08	
20.	9 − 0.4	
21.	9 − 0.04	
22.	9 − 0.004	

23.	9.9 − 5	
24.	9.9 − 0.5	
25.	0.99 − 0.5	
26.	0.99 − 0.05	
27.	4.7 − 2	
28.	4.7 − 0.2	
29.	0.47 − 0.2	
30.	0.47 − 0.02	
31.	8.4 − 1	
32.	8.4 − 0.1	
33.	0.84 − 0.1	
34.	7.2 − 5	
35.	7.2 − 0.5	
36.	0.72 − 0.5	
37.	0.72 − 0.05	
38.	8.6 − 7	
39.	8.6 − 0.7	
40.	0.86 − 0.7	
41.	0.86 − 0.07	
42.	5.1 − 4	
43.	5.1 − 0.4	
44.	0.51 − 0.4	

Number Correct: _____

Improvement: _____

Subtraction of Decimals—Round 2

Directions: Evaluate each expression.

1.	66 – 60	
2.	66 – 6	
3.	6.6 – 6	
4.	6.6 – 0.6	
5.	99 – 90	
6.	99 – 9	
7.	9.9 – 9	
8.	9.9 – 0.9	
9.	22 – 20	
10.	22 – 2	
11.	2.2 – 2	
12.	3 – 0.4	
13.	3 – 0.04	
14.	3 – 0.004	
15.	0.3 – 0.04	
16.	8 – 0.2	
17.	8 – 0.02	
18.	8 – 0.002	
19.	0.8 – 0.02	
20.	5 – 0.1	
21.	5 – 0.01	
22.	5 – 0.001	

23.	6.8 – 4	
24.	6.8 – 0.4	
25.	0.68 – 0.4	
26.	0.68 – 0.04	
27.	7.3 – 1	
28.	7.3 – 0.1	
29.	0.73 – 0.1	
30.	0.73 – 0.01	
31.	9.5 – 2	
32.	9.5 – 0.2	
33.	0.95 – 0.2	
34.	8.3 – 5	
35.	8.3 – 0.5	
36.	0.83 – 0.5	
37.	0.83 – 0.05	
38.	7.2 – 4	
39.	7.2 – 0.4	
40.	0.72 – 0.4	
41.	0.72 – 0.04	
42.	9.3 – 7	
43.	9.3 – 0.7	
44.	0.93 – 0.7	

Lesson 30: One-Step Problems in the Real World

355

Opening Exercise

Draw an example of each term, and write a brief description.

Acute

Obtuse

Right

Straight

Reflex

Example 1

$\angle ABC$ measures 90°. The angle has been separated into two angles. If one angle measures 57°, what is the measure of the other angle?

How are these two angles related?

What equation could we use to solve for x?

Now, let's solve.

Example 2

Michelle is designing a parking lot. She has determined that one of the angles should be 115°. What are the measures of angle $x°$ and angle $y°$?

How is angle $x°$ related to the 115° angle?

What equation would we use to show this?

How would you solve this equation?

How is angle $y°$ related to the angle that measures $115°$?

Example 3

A beam of light is reflected off a mirror. Below is a diagram of the reflected beam. Determine the missing angle measure.

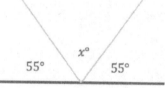

How are the angles in this question related?

What equation could we write to represent the situation?

How would you solve an equation like this?

Example 4

$\angle ABC$ measures 90°. It has been split into two angles, $\angle ABD$ and $\angle DBC$. The measure of the two angles is in a ratio of $4:1$. What is the measure of each angle?

Use a tape diagram to represent the ratio $4:1$.

What is the measure of each angle?

How can we represent this situation with an equation?

Solve the equation to determine the measure of each angle.

Exercises

Write and solve an equation in each of the problems.

1. $\angle ABC$ measures 90°. It has been split into two angles, $\angle ABD$ and $\angle DBC$. The measures of the two angles are in a ratio of $2:1$. What is the measure of each angle? Let $x°$ represent the measure of one of the unknown angles.

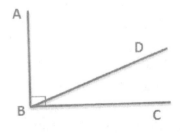

EUREKA MATH

2. Solve for $x°$.

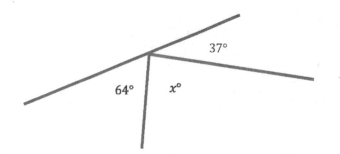

3. Candice is building a rectangular piece of a fence according to the plans her boss gave her. One of the angles is not labeled. Write an equation, and use it to determine the measure of the unknown angle.

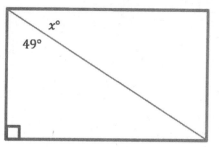

4. Rashid hit a hockey puck against the wall at a 38° angle. The puck hit the wall and traveled in a new direction. Determine the missing angle in the diagram.

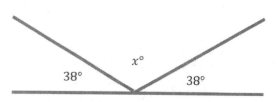

5. Jaxon is creating a mosaic design on a rectangular table. He has added two pieces to one of the corners. The first piece has an angle measuring 38° and is placed in the corner. A second piece has an angle measuring 27° and is also placed in the corner. Draw a diagram to model the situation. Then, write an equation, and use it to determine the measure of the unknown angle in a third piece that could be added to the corner of the table. Let $x°$ represent the measure of the unknown angle.

Name _____   Date _____

Write an equation, and solve for the missing angle in each question.

1. Alejandro is repairing a stained glass window. He needs to take it apart to repair it. Before taking it apart, he makes a sketch with angle measures to put it back together.

   Write an equation, and use it to determine the measure of the unknown angle.

   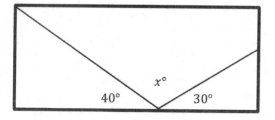

2. Hannah is putting in a tile floor. She needs to determine the angles that should be cut in the tiles to fit in the corner. The angle in the corner measures 90°. One piece of the tile will have a measure of 38°. Write an equation, and use it to determine the measure of the unknown angle.

   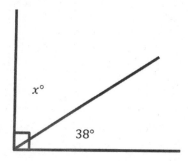

Angles: A Review

Students review the measures of four angles in this lesson. Using the information below, students create and solve one-step equations based on measurements of angles.

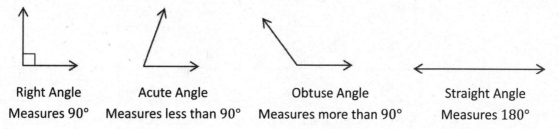

Right Angle	Acute Angle	Obtuse Angle	Straight Angle
Measures 90°	Measures less than 90°	Measures more than 90°	Measures 180°

Write and solve equations for each problem.

1. Solve for x.

> I know a right angle measures 90°. I also know that angle measurements are additive. I know part of the angle measure is 62°, but I don't know the value of x. If I add $x°$ and 62°, it will equal 90°.

$$x° + 62° = 90°$$
$$x° + 62° - 62° = 90° - 62°$$
$$x° = 28°$$

2. Solve for x.

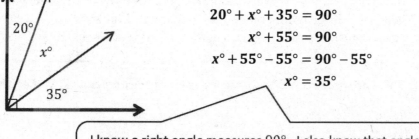

$$20° + x° + 35° = 90°$$
$$x° + 55° = 90°$$
$$x° + 55° - 55° = 90° - 55°$$
$$x° = 35°$$

> I know a right angle measures 90°. I also know that angle measurements are additive. I know part of the angle measure is 20°, and another part is 35°, but I don't know the value of x. If I add $x°$, 20°, and 35°, it will equal 90°.

3. Solve for x.

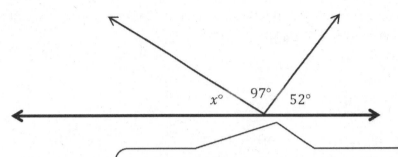

I know a straight angle measures $180°$. I also know that angle measurements are additive. I know part of the angle measure is $97°$, and another part is $52°$, but I don't know the value of x. If I add $x°$, $97°$, and $52°$, it will equal $180°$.

$$x° + 97° + 52° = 180°$$
$$x° + 149° = 180°$$
$$x° + 149° - 149° = 180° - 149°$$
$$x° = 31°$$

4. The measure of two angles have a sum of $90°$. The measures of the angels are in a ratio of $3:2$. Determine the measures of both angles.

$$3x° + 2x° = 90°$$
$$5x° = 90°$$
$$5x° \div 5 = 90° \div 5$$
$$x° = 18°$$

I know the ratio of the two unknown angles is $3:2$. This means there is a multiplicative comparison. The first angle is three times as many, and the second angle is two times as many. I can find the unknown angles by adding the three times as many, $3x$, and the two times as many, $2x$. Since angle measurements are additive, $3x° + 2x°$ is equal to $90°$.

If one x is $18°$, then two x's is $36°$, and three x's is $54°$. Since the ratio of the angles is $3:2$, then the angles measure $54°$ and $36°$.

Write and solve an equation for each problem.

1. Solve for $x°$.

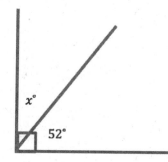

2. $\angle BAE$ measures 90°. Solve for $x°$.

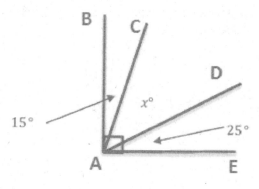

3. Thomas is putting in a tile floor. He needs to determine the angles that should be cut in the tiles to fit in the corner. The angle in the corner measures 90°. One piece of the tile will have a measure of 24°. Write an equation, and use it to determine the measure of the unknown angle. Let $x°$ represent the measure of the unknown angle.

4. Solve for $x°$.

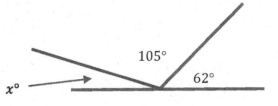

5. Aram has been studying the mathematics behind pinball machines. He made the following diagram of one of his observations. Determine the measure of the missing angle.

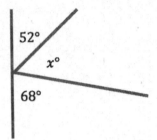

6. The measures of two angles have a sum of 90°. The measures of the angles are in a ratio of 2 : 1. Determine the measures of both angles. Let $x°$ represent the measure of one of the unknown angles.

7. The measures of two angles have a sum of 180°. The measures of the angles are in a ratio of 5 : 1. Determine the measures of both angles. Let $x°$ represent the measure of one of the unknown angles.

Example 1

Marcus reads for 30 minutes each night. He wants to determine the total number of minutes he will read over the course of a month. He wrote the equation $t = 30d$ to represent the total amount of time that he has spent reading, where t represents the total number of minutes read and d represents the number of days that he read during the month. Determine which variable is independent and which is dependent. Then, create a table to show how many minutes he has read in the first seven days.

Independent variable _____

Dependent variable _____

Example 2

Kira designs websites. She can create three different websites each week. Kira wants to create an equation that will give her the total number of websites she can design given the number of weeks she works. Determine the independent and dependent variables. Create a table to show the number of websites she can design over the first 5 weeks. Finally, write an equation to represent the number of websites she can design when given any number of weeks.

Independent variable _____

Dependent variable _____

Equation _____

Example 3

Priya streams movies through a company that charges her a $5 monthly fee plus $1.50 per movie. Determine the independent and dependent variables, write an equation to model the situation, and create a table to show the total cost per month given that she might stream between 4 and 10 movies in a month.

Independent variable _____

Dependent variable _____

Equation _____

Exercises

1. Sarah is purchasing pencils to share. Each package has 12 pencils. The equation $n = 12p$, where n is the total number of pencils and p is the number of packages, can be used to determine the total number of pencils Sarah purchased. Determine which variable is dependent and which is independent. Then, make a table showing the number of pencils purchased for 3–7 packages.

2. Charlotte reads 4 books each week. Let b be the number of books she reads each week, and let w be the number of weeks that she reads. Determine which variable is dependent and which is independent. Then, write an equation to model the situation, and make a table that shows the number of books read in under 6 weeks.

3. A miniature golf course has a special group rate. You can pay $20 plus $3 per person when you have a group of 5 or more friends. Let f be the number of friends and c be the total cost. Determine which variable is independent and which is dependent, and write an equation that models the situation. Then, make a table to show the cost for 5 to 12 friends.

4. Carlos is shopping for school supplies. He bought a pencil box for $3, and he also needs to buy notebooks. Each notebook is $2. Let t represent the total cost of the supplies and n be the number of notebooks Carlos buys. Determine which variable is independent and which is dependent, and write an equation that models the situation. Then, make a table to show the cost for 1 to 5 notebooks.

Name _____   Date _____

For each problem, determine the independent and dependent variables, write an equation to represent the situation, and then make a table with at least 5 values that models the situation.

1. Kyla spends 60 minutes of each day exercising. Let d be the number of days that Kyla exercises, and let m represent the total minutes of exercise in a given time frame. Show the relationship between the number of days that Kyla exercises and the total minutes that she exercises.

Independent variable _____

Dependent variable _____

Equation _____

2. A taxicab service charges a flat fee of $8 plus an additional $1.50 per mile. Show the relationship between the total cost and the number of miles driven.

Independent variable _____

Dependent variable _____

Equation _____

1. Barbara buys four books every month as part of a book club. To determine the number of books she can purchase in any given number of months, she uses the equation $b = 4m$, where b is the total number of books bought and m is the number of months. Name the independent variable and dependent variable. Create a table to show how many books she buys in less than 6 months.

The independent variable is m, or the number of months. The dependent variable, b, represents the number of books, and that depends on the number of months.

The independent variable is represented in the first column. I know that the number of months will change because the problem states *less than* 6 *months*. So, the number of months could be 1–5. Therefore, m, or the number of months, is the independent variable.

Number of Months (m)	Evaluating the Expression $b = 4m$	Total Number of Books (b)
1	$b = 4m$ $b = 4(1)$ $b = 4$	4
2	$b = 4m$ $b = 4(2)$ $b = 8$	8

The dependent variable is represented in the third column. The total number of books, b, depends on the number of months, so b is the dependent variable.

To determine the value of b, I replace m by the number of months it represents and then evaluate the expression.

Number of Months (m)	Total Amount of Books (b)
1	4
2	8
3	12
4	16
5	20

2. Tamara was given ten stamps. Each week, she collects three more stamps. Let w represent the number of weeks Tamara collects stamps and s represent the total number of stamps she has collected. Which variable is independent, and which is dependent? Write an equation to model the relationship, and make a table to show how many stamps she has from weeks 5–10.

$s = 3w + 10$. *The total number of stamps collected, s, is the dependent variable because it depends on the number of weeks Tamara collects stamps. The independent variable is the number of weeks Tamara collects stamps, w.* **10** *is a constant.*

The independent variable is represented in the first column. I know that the number of weeks will change because the problem states *from weeks 5–10.* Therefore, m, or the number of months, is the independent variable.

Number of Weeks(w)	Total Number of Stamps (s)
5	25
6	28
7	31
8	34
9	37
10	40

To determine the value of s, I replace w by the number of weeks it represents and then evaluate the equation.

$$s = 3w + 10$$
$$s = 3(5) + 10$$
$$s = 15 + 10$$
$$s = 25$$

I need to do this for all values of w.

Lesson 31:   Problems in Mathematical Terms

1. Jaziyah sells 3 houses each month. To determine the number of houses she can sell in any given number of months, she uses the equation $t = 3m$, where t is the total number of houses sold and m is the number of months. Name the independent and dependent variables. Then, create a table to show how many houses she sells in fewer than 6 months.

2. Joshua spends 25 minutes of each day reading. Let d be the number of days that he reads, and let m represent the total minutes of reading. Determine which variable is independent and which is dependent. Then, write an equation that models the situation. Make a table showing the number of minutes spent reading over 7 days.

3. Each package of hot dog buns contains 8 buns. Let p be the number of packages of hot dog buns and b be the total number of buns. Determine which variable is independent and which is dependent. Then, write an equation that models the situation, and make a table showing the number of hot dog buns in 3 to 8 packages.

4. Emma was given 5 seashells. Each week she collected 3 more. Let w be the number of weeks Emma collects seashells and s be the number of seashells she has total. Which variable is independent, and which is dependent? Write an equation to model the relationship, and make a table to show how many seashells she has from week 4 to week 10.

5. Emilia is shopping for fresh produce at a farmers' market. She bought a watermelon for $5, and she also wants to buy peppers. Each pepper is $0.75. Let t represent the total cost of the produce and n be the number of peppers bought. Determine which variable is independent and which is dependent, and write an equation that models the situation. Then, make a table to show the cost for 1 to 5 peppers.

6. A taxicab service charges a flat fee of $7 plus an additional $1.25 per mile driven. Show the relationship between the total cost and the number of miles driven. Which variable is independent, and which is dependent? Write an equation to model the relationship, and make a table to show the cost of 4 to 10 miles.

Opening Exercise

Xin is buying beverages for a party that come in packs of 8. Let p be the number of packages Xin buys and t be the total number of beverages. The equation $t = 8p$ can be used to calculate the total number of beverages when the number of packages is known. Determine the independent and dependent variables in this scenario. Then, make a table using whole number values of p less than 6.

Number of Packages (p)	Total Number of Beverages ($t = 8p$)
0	
1	
2	
3	
4	
5	

Example 1

Make a graph for the table in the Opening Exercise.

ample 2

Use the graph to determine which variable is the independent variable and which is the dependent variable. Then, state the relationship between the quantities represented by the variables.

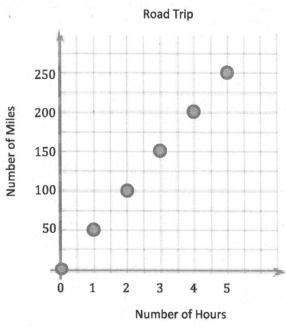

Road Trip

Exercises

1. Each week Quentin earns $30. If he saves this money, create a graph that shows the total amount of money Quentin has saved from week 1 through week 8. Write an equation that represents the relationship between the number of weeks that Quentin has saved his money, w, and the total amount of money in dollars that he has saved, s. Then, name the independent and dependent variables. Write a sentence that shows this relationship.

Lesson 32:   Multi-Step Problems in the Real World

2. Zoe is collecting books to donate. She started with 3 books and collects two more each week. She is using the equation $b = 2w + 3$, where b is the total number of books collected and w is the number of weeks she has been collecting books. Name the independent and dependent variables. Then, create a graph to represent how many books Zoe has collected when w is 5 or less.

3. Eliana plans to visit the fair. She must pay $5 to enter the fairgrounds and an additional $3 per ride. Write an equation to show the relationship between r, the number of rides, and t, the total cost in dollars. State which variable is dependent and which is independent. Then, create a graph that models the equation.

Name _____ Date _____

Determine which variable is the independent variable and which variable is the dependent variable. Write an equation, make a table, and plot the points from the table on the graph.

Enoch can type 40 words per minute. Let w be the number of words typed and m be the number of minutes spent typing.

Independent variable _____

Dependent variable _____

Equation _____

Beverly started saving money in a new account. She opened her account with $50. She adds $20 every week. Write an equation where w represents the number of weeks and t represents the total amount of money in the account, assuming no money is taken out and no interest is accrued. Determine which variable is independent and which is dependent. Then graph the total amount in the account for w being less than 8 weeks.

$$t = 20w + 50$$

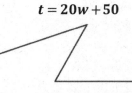

To determine the amount of money Beverly saves, I need to multiply the number of weeks by the amount of money she saves each week: $20w$. Then I need to add the original $50 to that product: $20w + 50$.

t is the dependent variable.

w is the independent variable.

I know that the number of weeks, w, is the independent variable. I am given those values: 0–7. I know that w will be measured along the x-axis. The total amount of money, t, depends on how many weeks Beverly saves. I know that t is the dependent variable and will be measured along the y-axis.

The independent variable is measured along the x-axis.

The dependent variable is measured along the y-axis.

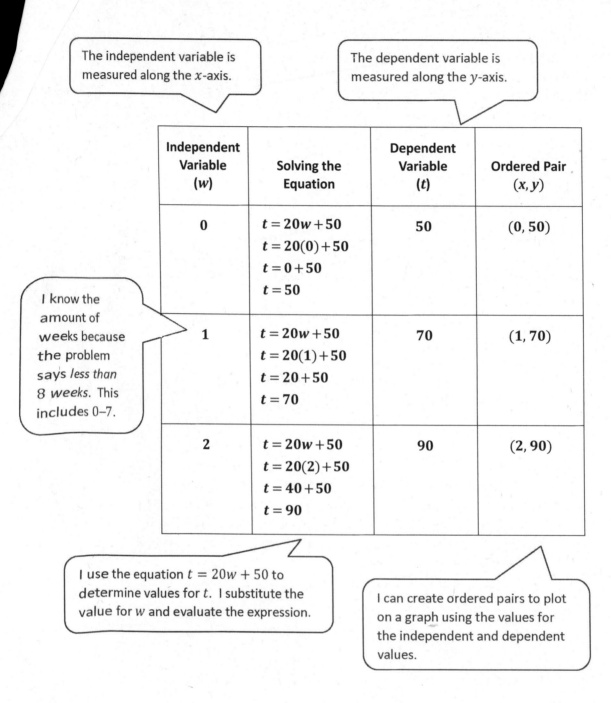

Independent Variable (w)	Solving the Equation	Dependent Variable (t)	Ordered Pair (x, y)
0	$t = 20w + 50$ $t = 20(0) + 50$ $t = 0 + 50$ $t = 50$	50	$(0, 50)$
1	$t = 20w + 50$ $t = 20(1) + 50$ $t = 20 + 50$ $t = 70$	70	$(1, 70)$
2	$t = 20w + 50$ $t = 20(2) + 50$ $t = 40 + 50$ $t = 90$	90	$(2, 90)$

I know the amount of weeks because the problem says *less than 8 weeks*. This includes 0–7.

I use the equation $t = 20w + 50$ to determine values for t. I substitute the value for w and evaluate the expression.

I can create ordered pairs to plot on a graph using the values for the independent and dependent values.

Number of Weeks (w)	Total Amount of Money in Dollars (t)
0	50
1	70
2	90
3	110
4	130
5	150
6	170
7	190

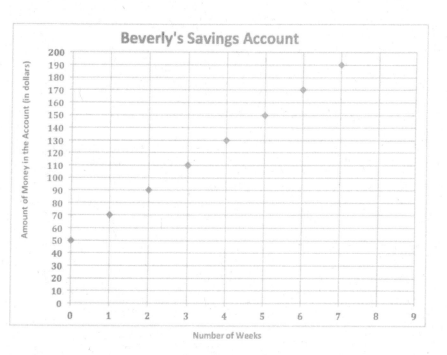

A point is placed at the intersection of the (x, y) coordinates. Beginning at zero, I move horizontally along the x-axis first to the first x-value and then vertically along the y-axis to the first y-value. There I place a point. I continue to do this for each of the ordered pairs created from the table to the left.

1. Caleb started saving money in a cookie jar. He started with $25. He adds $10 to the cookie jar each week. Write an equation where w is the number of weeks Caleb saves his money and t is the total amount in dollars in the cookie jar. Determine which variable is the independent variable and which is the dependent variable. Then, graph the total amount in the cookie jar for w being less than 6 weeks.

2. Kevin is taking a taxi from the airport to his home. There is a $6 flat fee for riding in the taxi. In addition, Kevin must also pay $1 per mile. Write an equation where m is the number of miles and t is the total cost in dollars of the taxi ride. Determine which variable is independent and which is dependent. Then, graph the total cost for m being less than 6 miles.

...ana started with $10. She saved an additional $5 each week. Write an equation that can be used to determine the total amount saved in dollars, t, after a given number of weeks, w. Determine which variable is independent and which is dependent. Then, graph the total amount saved for the first 8 weeks.

4. Aliyah is purchasing produce at the farmers' market. She plans to buy $10 worth of potatoes and some apples. The apples cost $1.50 per pound. Write an equation to show the total cost of the produce, where T is the total cost, in dollars, and a is the number of pounds of apples. Determine which variable is dependent and which is independent. Then, graph the equation on the coordinate plane.

Lesson 32:   Multi-Step Problems in the Real World

Example 1

What value(s) does the variable have to represent for the equation or inequality to result in a true number sentence?
What value(s) does the variable have to represent for the equation or inequality to result in a false number sentence?

  a. $y + 6 = 16$

  b. $y + 6 > 16$

  c. $y + 6 \geq 16$

  d. $3g = 15$

  e. $3g < 15$

  f. $3g \leq 15$

le 2

ch of the following number(s), if any, make the equation or inequality true: $\{0, 3, 5, 8, 10, 14\}$?

a. $m + 4 = 12$

b. $m + 4 < 12$

c. $f - 4 = 2$

d. $f - 4 > 2$

e. $\dfrac{1}{2}h = 8$

f. $\dfrac{1}{2}h \geq 8$

Lesson 33:   From Equations to Inequalities

Exercises

Choose the number(s), if any, that make the equation or inequality true from the following set of numbers: $\{0, 1, 5, 8, 11, 17\}$.

1. $m + 5 = 6$

2. $m + 5 \leq 6$

3. $5h = 40$

4. $5h > 40$

5. $\dfrac{1}{2}y = 5$

6. $\dfrac{1}{2}y \leq 5$

7. $k - 3 = 20$

8. $k - 3 > 20$

Name _____   Date _____

Choose the number(s), if any, that make the equation or inequality true from the following set of numbers:
{3, 4, 7, 9, 12, 18, 32}.

1. $\frac{1}{3}f = 4$

2. $\frac{1}{3}f < 4$

3. $m + 7 = 20$

4. $m + 7 \geq 20$

Moving from Equations to Inequalities

Students move from naming the values that make the sentence true or false to using a set of numbers and determining whether or not the numbers in the set make the equation or inequality true or false.

Choose the numbers that make the equation or inequality true from the following set of numbers: $\{2, 4, 6, 8, 9, 17\}$

1. $m - 2 = 6$

   $\{8\}$

   8 is the only number that makes this equation true.
   $$m - 2 + 2 = 6 + 2$$
   $$m = 8$$

2. $m - 2 < 6$

   $\{2, 4, 6\}$

   $$m - 2 + 2 < 6 + 2$$
   $$m < 8$$
   Because the number that m represents has to be less than 8, the only numbers from the set that are less than 8 are 2, 4, and 6.

3. $3x = 27$

   $\{9\}$

   9 is the only number that makes this equation true.
   $$3x \div 3 = 27 \div 3$$
   $$x = 9$$

4. $3x \geq 27$

   $\{9, 17\}$

   $$3x \div 3 \geq 27 \div 3$$
   $$x \geq 9$$
   Because the number that x represents has to be greater than or equal to 9, the only numbers from the set that are greater than or equal to 9 are 9 and 17.

$= 6$

There is no number in the set that makes this equation true.

$$\frac{1}{5}h \cdot 5 = 6 \cdot 5$$

$$h = 30$$

Because the number that h represents has to be equal to 30, and none of the number choices from the set are 30, then there is no number in the set that makes this equation true.

Choose the number(s), if any, that make the equation or inequality true from the following set of numbers: {0, 3, 4, 5, 9, 13, 18, 24}.

1. $h - 8 = 5$

2. $h - 8 < 5$

3. $4g = 36$

4. $4g \geq 36$

5. $\dfrac{1}{4}y = 7$

6. $\dfrac{1}{4}y > 7$

7. $m - 3 = 10$

8. $m - 3 \leq 10$

Example 1

Statement	Inequality	Graph
a. Caleb has at least $5.	_____	2 3 4 5 6 7 8
b. Tarek has more than $5.	_____	2 3 4 5 6 7 8
c. Vanessa has at most $5.	_____	2 3 4 5 6 7 8
d. Li Chen has less than $5.	_____	2 3 4 5 6 7 8

Example 2

Kelly works for Quick Oil Change. If customers have to wait longer than 20 minutes for the oil change, the company does not charge for the service. The fastest oil change that Kelly has ever done took 6 minutes. Show the possible customer wait times in which the company charges the customer.

4   6   8   10   12   14   16   18   20   22   24

Example 3

Gurnaz has been mowing lawns to save money for a concert. Gurnaz will need to work for at least six hours to save enough money, but he must work fewer than 16 hours this week. Write an inequality to represent this situation, and then graph the solution.

2   4   6   8   10   12   14   16   18   20   22

Exercises 1–5

Write an inequality to represent each situation. Then, graph the solution.

1. Blayton is at most 2 meters above sea level.

-1 0 1 2 3 4 5

2. Edith must read for a minimum of 20 minutes.

17 18 19 20 21 22 23

3. Travis milks his cows each morning. He has never gotten fewer than 3 gallons of milk; however, he always gets fewer than 9 gallons of milk.

2 3 4 5 6 7 8 9 10

4. Rita can make 8 cakes for a bakery each day. So far, she has orders for more than 32 cakes. Right now, Rita needs more than four days to make all 32 cakes.

2 3 4 5 6 7 8 9 10

5. Rita must have all the orders placed right now done in 7 days or fewer. How will this change your inequality and your graph?

2 3 4 5 6 7 8 9 10

Lesson 34: Writing and Graphing Inequalities in Real-World Problems

© 2019 Great Minds®. eureka-math.org

EUREKA MATH

Possible Extension Exercises 6–10

6. Kasey has been mowing lawns to save up money for a concert. He earns $15 per hour and needs at least $90 to go to the concert. How many hours should he mow?

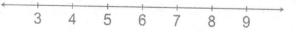

7. Rachel can make 8 cakes for a bakery each day. So far, she has orders for more than 32 cakes. How many days will it take her to complete the orders?

8. Ranger saves $70 each week. He needs to save at least $2,800 to go on a trip to Europe. How many weeks will he need to save?

9. Clara has less than $75. She wants to buy 3 pairs of shoes. What price shoes can Clara afford if all the shoes are the same price?

10. A gym charges $25 per month plus $4 extra to swim in the pool for an hour. If a member only has $45 to spend each month, at most how many hours can the member swim?

Exercises 1–5

Write an inequality to represent each situation. Then, graph the solution.

1. Blayton is at most 2 meters above sea level.

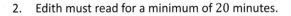

2. Edith must read for a minimum of 20 minutes.

3. Travis milks his cows each morning. He has never gotten fewer than 3 gallons of milk; however, he always gets fewer than 9 gallons of milk.

4. Rita can make 8 cakes for a bakery each day. So far, she has orders for more than 32 cakes. Right now, Rita needs more than four days to make all 32 cakes.

5. Rita must have all the orders placed right now done in 7 days or fewer. How will this change your inequality and your graph?

Lesson 34:   Writing and Graphing Inequalities in Real-World Problems

EUREKA MATH

Example 1

Statement	Inequality	Graph

a. Caleb has at least $5. _____

2 3 4 5 6 7 8

b. Tarek has more than $5. _____

2 3 4 5 6 7 8

c. Vanessa has at most $5. _____

2 3 4 5 6 7 8

d. Li Chen has less than $5. _____

2 3 4 5 6 7 8

Example 2

Kelly works for Quick Oil Change. If customers have to wait longer than 20 minutes for the oil change, the company does not charge for the service. The fastest oil change that Kelly has ever done took 6 minutes. Show the possible customer wait times in which the company charges the customer.

4 6 8 10 12 14 16 18 20 22 24

Example 3

Gurnaz has been mowing lawns to save money for a concert. Gurnaz will need to work for at least six hours to save enough money, but he must work fewer than 16 hours this week. Write an inequality to represent this situation, and then graph the solution.

2 4 6 8 10 12 14 16 18 20 22

Choose the number(s), if any, that make the equation or inequality true from the following set of numbers: {0, 3, 4, 5, 9, 13, 18, 24}.

1. $h - 8 = 5$

2. $h - 8 < 5$

3. $4g = 36$

4. $4g \geq 36$

5. $\frac{1}{4}y = 7$

6. $\frac{1}{4}y > 7$

7. $m - 3 = 10$

8. $m - 3 \leq 10$

5. $\frac{1}{5}h = 6$

There is no number in the set that makes this equation true.

$$\frac{1}{5}h \cdot 5 = 6 \cdot 5$$
$$h = 30$$

Because the number that h represents has to be equal to 30, and none of the number choices from the set are 30, then there is no number in the set that makes this equation true.

Moving from Equations to Inequalities

Students move from naming the values that make the sentence true or false to using a set of numbers and determining whether or not the numbers in the set make the equation or inequality true or false.

Choose the numbers that make the equation or inequality true from the following set of numbers: $\{2, 4, 6, 8, 9, 17\}$

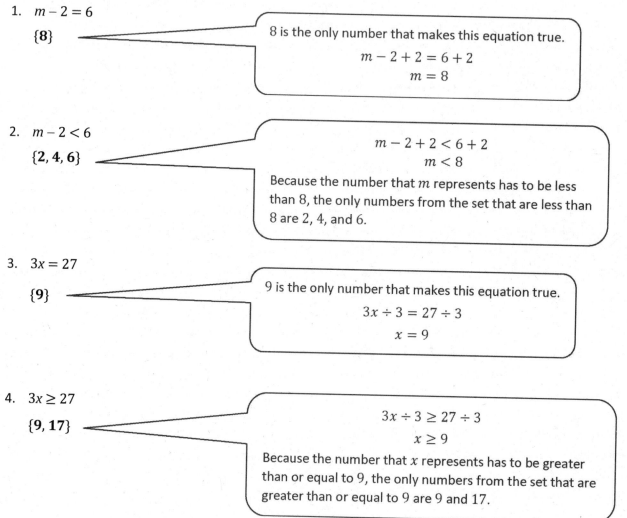

1. $m - 2 = 6$

   $\{8\}$

   8 is the only number that makes this equation true.
   $$m - 2 + 2 = 6 + 2$$
   $$m = 8$$

2. $m - 2 < 6$

   $\{2, 4, 6\}$

   $$m - 2 + 2 < 6 + 2$$
   $$m < 8$$
   Because the number that m represents has to be less than 8, the only numbers from the set that are less than 8 are 2, 4, and 6.

3. $3x = 27$

   $\{9\}$

   9 is the only number that makes this equation true.
   $$3x \div 3 = 27 \div 3$$
   $$x = 9$$

4. $3x \geq 27$

   $\{9, 17\}$

   $$3x \div 3 \geq 27 \div 3$$
   $$x \geq 9$$
   Because the number that x represents has to be greater than or equal to 9, the only numbers from the set that are greater than or equal to 9 are 9 and 17.

Name _____   Date _____

For each question, write an inequality. Then, graph your solution.

1. Keisha needs to make at least 28 costumes for the school play. Since she can make 4 costumes each week, Keisha plans to work on the costumes for at least 7 weeks.

2. If Keisha has to have the costumes complete in 10 weeks or fewer, how will our solution change?

Graphing Inequalities

When an inequality has a variable that is less than or equal to or greater than or equal to ($\leq$ or $\geq$) a number, then (because the solution includes the number) the point is plotted on the graph. For example: $x \leq 9$ (x is less than or equal to 9). This solution will include 9 and all numbers less than 9. To plot 9 on the graph, it is represented with a closed circle because it is a solution to the inequality. A ray to the left of 9 represents all rational numbers less than 9 because they are all solutions to the inequality.

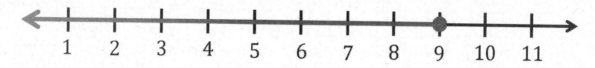

When an inequality is less than or greater than a number ($<$ or $>$), the solution does not include the number. The number is the beginning place, and instead of plotting a closed point on the graph, an open point (or open circle) determines the beginning place. For example: $x > 9$ (x is greater than 9). An open point at 9 is plotted on the graph since 9 is not a solution to the inequality but a beginning point. A ray to the right of 9 represents all rational numbers greater than 9 because they are all solutions to the inequality.

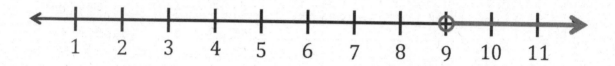

Write and graph an inequality for each problem.

1. At least 64

   $x \geq 64$

   I know numbers that are *at least* 64 include 64 as the least amount and any number greater than 64. I need to plot 64 with a point and all numbers greater than 64 with a ray to the right of 64.

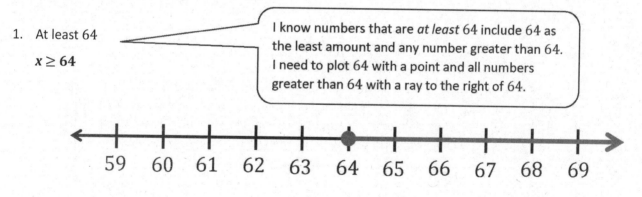

2. Less than 10

$x < 10$

I know numbers that are *less than* 10 do not include 10, but I should start with 10 as a beginning point. I need to plot an open point (or circle) at 10 and represent all numbers less than 10 with a ray to the left of 10.

3. Cameron needs at least 10 minutes to finish his assignment. However, he must finish in under 30 minutes.

$10 \leq x < 30$

I know numbers that are *at least* 10 include 10 as the least amount. I should start with 10 as a beginning point. If Cameron needs to finish *under* 30 minutes, 30 is my stopping point. The solution set does not include 30 because *under* 30 *is less than* 30.

All numbers between 10 (including 10) and 30 (not including 30) need to be represented. I need to plot a point at 10 and represent all numbers greater than 10 with a line segment to the right of 10 until I reach 30. My stopping point is 30, which I will represent with an open point, or an open circle.

Lesson 34:   Writing and Graphing Inequalities in Real-World Problems

EUREKA MATH®

Write and graph an inequality for each problem.

1. At least 13

```
 ←—+——+——+——+——+——+——+——+——+——+——+——→
   10  11  12  13  14  15  16  17  18  19  20
```

2. Less than 7

```
 ←—+——+——+——+——+——+——+——+——+——→
    2   3   4   5   6   7   8   9   10
```

3. Chad will need at least 24 minutes to complete the 5K race. However, he wants to finish in under 30 minutes.

```
 ←——+————+————+————+————+————+——→
    22   24   26   28   30   32
```

4. Eva saves $60 each week. Since she needs to save at least $2,400 to go on a trip to Europe, she will need to save for at least 40 weeks.

```
 ←——+————+————+————+————+————+————+——→
    25   30   35   40   45   50   55
```

5. Clara has $100. She wants to buy 4 pairs of the same pants. Due to tax, Clara can afford pants that are less than $25.

```
 |——+——+——+——+——+——+——+——+——+——+——→
   0   5   10  15  20  25  30  35  40  45  50
```

6. A gym charges $30 per month plus $4 extra to swim in the pool for an hour. Because a member has just $50 to spend at the gym each month, the member can swim at most 5 hours.

```
 |——+——+——+——+——+——+——→
   0   1   2   3   4   5   6
```

Credits

Great Minds® has made every effort to obtain permission for the reprinting of all copyrighted material. If any owner of copyrighted material is not acknowledged herein, please contact Great Minds for proper acknowledgment in all future editions and reprints of this module.